Research Notes in Mathematics

Submission of proposals for consideration
Suggestions for publication, in the form of outlines and representative samples, are invited by the Editorial Board for assessment. Intending authors should approach one of the main editors or another member of the Editorial Board, citing the relevant AMS subject classifications. Alternatively, outlines may be sent directly to one of the publisher's offices. Refereeing is by members of the board and other mathematical authorities in the topic concerned, throughout the world.

Preparation of accepted manuscripts
On acceptance of a proposal, the publisher will supply full instructions for the preparation of manuscripts in a form suitable for direct photo-lithographic reproduction. Specially printed grid sheets are provided and a contribution is offered by the publisher towards the cost of typing. Word processor output, subject to the publisher's approval, is also acceptable.

Illustrations should be prepared by the authors, ready for direct reproduction without further improvement. The use of hand-drawn symbols should be avoided wherever possible, in order to maintain maximum clarity of the text.

The publisher will be pleased to give any guidance necessary during the preparation of a typescript, and will be happy to answer any queries.

Important note
In order to avoid later retyping, intending authors are strongly urged not to begin final preparation of a typescript before receiving the publisher's guidelines and special paper. In this way it is hoped to preserve the uniform appearance of the series.

Advanced Publishing Program
Pitman Publishing Inc
1020 Plain Street
Marshfield, MA 02050, USA
(tel (617) 837 1331)

Advanced Publishing Program
Pitman Publishing Limited
128 Long Acre
London WC2E 9AN, UK
(tel 01-379 7383)

Titles in this series

56 An approach to algebraic K-theory
A J Berrick
57 Convex analysis and optimization
J-P Aubin and R B Vintner
58 Convex analysis with applications in the differentiation of convex functions
J R Giles
59 Weak and variational methods for moving boundary problems
C M Elliott and J R Ockendon
60 Nonlinear partial differential equations and their applications: Collège de France Seminar. Volume II
H Brezis and J L Lions
61 Singular systems of differential equations II
S L Campbell
62 Rates of convergence in the central limit theorem
Peter Hall
63 Solution of differential equations by means of one-parameter groups
J M Hill
64 Hankel operators on Hilbert space
S C Power
65 Schrödinger-type operators with continuous spectra
M S P Eastham and H Kalf
66 Recent applications of generalized inverses
S L Campbell
67 Riesz and Fredholm theory in Banach algebra
B A Barnes, G J Murphy, M R F Smyth and T T West
68 Evolution equations and their applications
F Kappel and W Schappacher
69 Generalized solutions of Hamilton-Jacobi equations
P L Lions
70 Nonlinear partial differential equations and their applications: Collège de France Seminar. Volume III
H Brezis and J L Lions
71 Spectral theory and wave operators for the Schrödinger equation
A M Berthier
72 Approximation of Hilbert space operators I
D A Herrero
73 Vector valued Nevanlinna Theory
H J W Ziegler
74 Instability, nonexistence and weighted energy methods in fluid dynamics and related theories
B Straughan
75 Local bifurcation and symmetry
A Vanderbauwhede
76 Clifford analysis
F Brackx, R Delanghe and F Sommen
77 Nonlinear equivalence, reduction of PDEs to ODEs and fast convergent numerical methods
E E Rosinger
78 Free boundary problems, theory and applications. Volume I
A Fasano and M Primicerio
79 Free boundary problems, theory and applications. Volume II
A Fasano and M Primicerio
80 Symplectic geometry
A Crumeyrolle and J Grifone
81 An algorithmic analysis of a communication model with retransmission of flawed messages
D M Lucantoni
82 Geometric games and their applications
W H Ruckle
83 Additive groups of rings
S Feigelstock
84 Nonlinear partial differential equations and their applications: Collège de France Seminar. Volume IV
H Brezis and J L Lions
85 Multiplicative functionals on topological algebras
T Husain
86 Hamilton-Jacobi equations in Hilbert spaces
V Barbu and G Da Prato
87 Harmonic maps with symmetry, harmonic morphisms and deformations of metrics
P Baird
88 Similarity solutions of nonlinear partial differential equations
L Dresner
89 Contributions to nonlinear partial differential equations
C Bardos, A Damlamian, J I Díaz and J Hernández
90 Banach and Hilbert spaces of vector-valued functions
J Burbea and P Masani
91 Control and observation of neutral systems
D Salamon
92 Banach bundles, Banach modules and automorphisms of C*-algebras
M J Dupré and R M Gillette
93 Nonlinear partial differential equations and their applications: Collège de France Seminar. Volume V
H Brezis and J L Lions
94 Computer algebra in applied mathematics: an introduction to MACSYMA
R H Rand
95 Advances in nonlinear waves. Volume I
L Debnath
96 FC-groups
M J Tomkinson
97 Topics in relaxation and ellipsoidal methods
M Akgül
98 Analogue of the group algebra for topological semigroups
H Dzinotyiweyi
99 Stochastic functional differential equations
S E A Mohammed
100 Optimal control of variational inequalities
V Barbu
101 Partial differential equations and dynamical systems
W E Fitzgibbon III
102 Approximation of Hilbert space operators. Volume II
C Apostol, L A Fialkow, D A Herrero and D Voiculescu
103 Nondiscrete induction and iterative processes
V Ptak and F-A Potra
104 Analytic functions – growth aspects
O P Juneja and G P Kapoor
105 Theory of Tikhonov regularization for Fredholm equations of the first kind
C W Groetsch

106 Nonlinear partial differential equations and free boundaries. Volume I
J I Díaz

107 Tight and taut immersions of manifolds
T E Cecil and P J Ryan

108 A layering method for viscous, incompressible L_p flows occupying R^n
A Douglis and E B Fabes

109 Nonlinear partial differential equations and their applications: Collège de France Seminar. Volume VI
H Brezis and J L Lions

110 Finite generalized quadrangles
S E Payne and J A Thas

111 Advances in nonlinear waves. Volume II
L Debnath

112 Topics in several complex variables
E Ramírez de Arellano and D Sundararaman

113 Differential equations, flow invariance and applications
N H Pavel

114 Geometrical combinatorics
F C Holroyd and R J Wilson

115 Generators of strongly continuous semigroups
J A van Casteren

116 Growth of algebras and Gelfand–Kirillov dimension
G R Krause and T H Lenagan

117 Theory of bases and cones
P K Kamthan and M Gupta

118 Linear groups and permutations
A R Camina and E A Whelan

119 General Wiener–Hopf factorization methods
F-O Speck

120 Free boundary problems: applications and theory, Volume III
A Bossavit, A Damlamian and M Fremond

121 Free boundary problems: applications and theory, Volume IV
A Bossavit, A Damlamian and M Fremond

122 Nonlinear partial differential equations and their applications: Collège de France Seminar. Volume VII
H Brezis and J. L. Lions

123 Geometric methods in operator algebras
H Araki and E G Effros

124 Infinite dimensional analysis–stochastic processes
S Albeverio

125 Ennio de Giorgi Colloquium
P Krée

126 Almost-periodic functions in abstract spaces
S Zaidman

127 Nonlinear variational problems
A Marino, L Modica, S Spagnolo and M Degiovanni

128 Second-order systems of partial differential equations in the plane
L K Hua, W Lin and C-Q Wu

129 Asymptotics of high-order ordinary differential equations
R B Paris and A D Wood

130 Stochastic differential equations
R Wu

Linear groups and permutations

A R Camina & E A Whelan

University of East Anglia

Linear groups and permutations

Pitman Advanced Publishing Program
BOSTON · LONDON · MELBOURNE

PITMAN PUBLISHING INC
1020 Plain Street, Marshfield, Massachusetts 02050

PITMAN PUBLISHING LIMITED
128 Long Acre, London WC2E 9AN

Associated Companies
Pitman Publishing Pty Ltd, Melbourne
Pitman Publishing New Zealand Ltd, Wellington
Copp Clark Pitman, Toronto

First published 1985

AMS Subject Classifications: (main) 20B10, 20C15, 20B25
(subsidiary) 13XX, 51AXX, 51EXX

ISSN 0743-0337

Library of Congress Cataloging in Publication Data

Camina, A. R.
Linear groups and permutations.

Bibliography: p.
Includes index.
1. Permutation groups. 2. Representations of groups. 3. Automorphisms. I. Whelan, E. A. II. Title.
QA171.C258 1985 512′.22 84-25388
ISBN 0-273-08672-3

British Library Cataloguing in Publication Data

Camina, A.R.
Linear groups and permutations.—(Research notes in mathematics, ISSN 0743-0337; 118)
1. Finite groups
I. Title II. Whelan, E.A. III. Series
512′.2 QA171

ISBN 0-273-08672-3

Reproduced and printed by photolithography
in Great Britain by Biddles Ltd, Guildford

Contents

Introduction

We have set ourselves the task in these notes of explaining the work of C. Hering on linear groups over finite fields. These results have an important role to play especially now that the classification of the finite simple groups has been completed. We should not however underestimate their significance in the study of finite soluble groups. The results of Hering as well as a discussion of their applications are the subject of Chapter 5. This chapter also contains an introduction to the theory of block designs (incidence structures).

Our approach to this particular topic is typical of our whole attitude: to present the results in a technically efficient manner without sacrificing either generality or the need to see the work in a more abstract setting. The material is based on an M. Sc. course given by the first author at the University of East Anglia. As a consequence it is hoped that the book will be accessible to most students recently graduated in mathematics. However we hope that there will be much of interest to the experienced algebraist, whether they be ring theorist, group theorist or number theorist.

The ring theory has been presented in an unusual manner and this has affected the way that the representation theory has been tackled. We have introduced the term "ringlet". Whilst accepting that new terminology is usually unwelcome we feel that the confusion caused by the inconsistency between the various definitions of "ring" warrants this decision. The categories of rings with 1 and of rings without 1 are sufficiently different to justify the new nomenclature.

There is one area where we have taken a rather careful approach for a book of this type, namely, in the theorem about the change of field for group presentations. The fact that the group algebra of a finite group over a splitting field of coprime characteristic is well-behaved is well-known, and not particularly deep. The result that the degrees of the simple modules are independent of the characteristic (given the coprime condition) is a much deeper result. We feel that many of the less encyclopaedic texts tend to fumble some of the details, possibly to avoid the essential number theory which is involved. We have tried to give a complete but readable account of this theorem.

Finally we would like to thank Graham Everest for his helpful comments, Carol Haines for her typing and Bridget Buckley of Pitman's for her patience and support.

A. R. Camina

E. A. Whelan

Conventions

"Ringlet" means associative but not necessarily commutative ring, not necessarily with unity element. "Ring" means ringlet with unity element, 1. "Ideal" means two-sided ideal unless otherwise qualified, and the characterisations "left", "right", "maximal", "maximal left", "prime" applied to one- or two-sided ideals have the usual meanings.

If H, G are groups then

$H \leq G$ means "H is a subgroup of G"

$H < G$ means "$H \leq G$ and $H \neq G$"

$H \trianglelefteq G$ means "H is a normal subgroup of G"

$H \triangleleft G$ means "$H \trianglelefteq G$ and $H \neq G$".

The symbol $\subset$ denotes strict set-theoretic inclusion.

References in square brackets such as [3] are to the bibliography at the end of the volume. Numbers in round brackets such as (1.1) refer to the item of that number in these notes.

The symbols $\mathbb{N}$, $\mathbb{Z}$, $\mathbb{Q}$, $\mathbb{R}$, $\mathbb{C}$ denote respectively the natural numbers, the rational integers, the field of rational numbers, the real field, and the complex field.

The abbreviations "a. c. c." and "d. c. c." stand for "ascending chain condition" and "descending chain condition" respectively.

If $a, b \in \mathbb{Z}$ then (a, b) usually denotes the non-negative generator of the ideal $a\mathbb{Z} + b\mathbb{Z}$, that is, the highest common factor (h. c. f.) of a and b.

If G is a group and $g \in G$ then $\langle g \rangle$ denotes the subgroup generated by G. If G is additive abelian then $G^{\#} = G - \{0\}$.

Mappings (except of one-sided modules) are generally written on the right. There are a few exceptions to this guideline, generally where tradition hallows a different usage. Morphisms of one-sided modules are written on the opposite side to scalars.

There is one theorem in module theory which for reasons of space we have not proved, but on which we rely heavily: the decomposition theorem for finitely generated modules over a commutative principal ideal domain. This result, or similar ones, can be found in good undergraduate algebra text books, some of which we list as background reading with our references. For our purposes, it is sufficient to know that a (one-sided) module over such a ring decomposes into a free direct summand and a direct summand with non-zero annihilator, provided that it is finitely generated.

1 Basic group theory

INTRODUCTION

The chief aims of this chapter are to introduce and explain several fundamental group-theoretic ideas which recur throughout the book, and to develop the basic results about them. We shall also prove some important technical group-theoretic lemmas which will be needed later.

The main ideas which are expounded here concern permutation groups, nilpotent groups, the Fitting and the Frattini subgroups of a group. Readers familiar with these topics may nevertheless find something of interest in our coverage, since we have sought to state and to prove each result in the greatest useful generality. Notably, we avoid where it is avoidable, the assumption that we are dealing with finite groups. It turns out that this policy can be carried through fairly thoroughly, and that the results thus obtained both are interesting in themselves and facilitate extensions of known ring-theoretic and group theoretic results in later chapters.

PERMUTATION GROUPS

Strictly speaking, a permutation group is simply a subgroup G of the group S^{Ω} of permutations of a non-empty set Ω. In accordance with the policy outlined in our opening statement of conventions, we denote the image of x in Ω under the action of g in G by xg. In the obvious way G also acts as a permutation group on the set of subsets of Ω, with $\Delta g = \{\delta g : \delta \in \Delta\}$ for an element g of G and $\emptyset \neq \Delta \subseteq \Omega$, and $\emptyset g = \emptyset$ for $g \in G$. If Δ is a non-empty subset of Ω then Δ is said to be fixed by G if $\Delta g = g$ for each g in G (rather as one would expect). If such a Δ is fixed by G but no non-trivial subset of Δ is fixed by G then Δ is called an orbit of G (or, an orbit of Ω under G). Alternatively, if we define a relation $\sim$ on Ω by $x \sim y$ if and only if $xg = y$ for some g in G, it is easily verified (Exercise!) that $\sim$ is an equivalence relation on Ω, and that the equivalence classes of $\sim$ are precisely the orbits of Ω under G. If Ω itself is an orbit of G (and hence the only one) then G is said to be transitive. If α is a point of Ω, the orbit containing α is also referred to as the orbit of α.

If G is a permutation group on Ω and Δ is a non-empty subset of Ω, we define the stabiliser of Δ to be $G_\Delta = \{g \in G : \delta g = \delta \,\forall \delta \in \Delta\}$. This group is sometimes called the pointwise stabiliser of Δ to distinguish it from the larger group $\{h \in G : \Delta h = \Delta\}$, with which we are not much concerned here. In the case $\Delta = \{\alpha\}$ is a singleton, we write $G_\Delta = G_\alpha$. If Δ is an orbit of G and α is in Δ then there is a bijection between the (right, or left) cosets of G_α and the points of Δ. For example, if the right cosets of G_α are $\{G_\alpha x_i : i \in I\}$ and Δ is the orbit of α, then $\Delta = \{\alpha x_i : i \in I\}$, where $i, j \in I$ and $i \neq j$ is easily seen to imply $\alpha x_i \neq \alpha x_j$. When Δ is finite (in particular when G itself is finite) we have $|\Delta| = [G : G_\alpha]$. If G is a permutation group on Ω and H is a subgroup of G then H is also a permutation group on Ω.

If G is an abstract group and θ is a homomorphism of G into the group S^Ω of permutations of a non-empty set S, we say that the ordered pair (G, θ) is a representation of G as a permutation group on Ω. Sometimes we will refer rather loosely to θ as being the representation. On the other hand, if θ is a monomorphism then the representation is called faithful, and in that case we often focus on G rather than θ, and speak of G as being (rather than having a representation as) a permutation group on Ω: this is sometimes called using θ to identify G with its image in S^Ω. These slightly loose usages can be objected to, but they are hallowed by tradition and by utility, so we will follow the general fashion and use terminology and emphasis as suits the purpose at hand in a given context. It is important for the reader, however, to be aware that there is a problem here and that from time to time it will be essential to keep very careful track of the distinctions between G, θ and $G\theta \subseteq S^\Omega$.

1.1 Proposition. Let G be a transitive permutation group on Ω and N be a subgroup of G which is transitive on Ω. Then for any α in Ω we have $G = G_\alpha N$. If in addition N is normal in G then $G = G_\alpha N = NG_\alpha$.

Proof. Let $g \in G$ and $\alpha \in \Omega$. Since N is transitive on Ω there exists n in N such that $\alpha g = \alpha n$. Thus $gn^{-1} \in G_\alpha$ and so $g \in G_\alpha N$. Hence $G = G_\alpha N$, and the final assertion follows by looking at cosets. //

Suppose that N is a finite normal subgroup of a group G, p is a prime dividing $|N|$, and Ω is the set of Sylow p-subgroups of N. Then under the natural action of conjugation G has a representation as a permutation group on Ω, and the conditions of (1.1) are satisfied, so we obtain the following corollary, known as the "Frattini argument".

1.2 Corollary. Let N be a finite normal subgroup of a group G and let p be a Sylow p-subgroup of N for some prime p dividing the order of N, then $G = N_G(P)N = NN_G(P)$. //

A regular permutation group on Ω is a permutation group G on a set Ω such that $G_\alpha = 1$ for each α in Ω.

1.3 Lemma. Let G be a transitive permutation group on Ω, and assume there is an α in Ω such that the group G_α contains a subgroup N which is normal in G. Then $N = 1$.

Proof. Let $g \in N$. If $\beta \in \Omega$ then by transitivity there exists $h \in G$ such that $\alpha h = \beta$. By normality, $hg = g'h$ for some $g' \in N \subseteq G_\alpha$. Hence $\beta g = \alpha hg = \alpha g'h = \alpha h = \beta$, so g acts as the identity on Ω. //

1.4 Corollary. Let G be a transitive permutation group on Ω, and assume that for each point α of Ω the group G_α is normal in G (for example, if G is abelian). Then G is a regular permutation group on Ω. //

1.5 Lemma. Let G be a transitive permutation group on Ω and let N be a normal subgroup regular as a permutation group on Ω. Given any point α of Ω and g in G, then $g = nh$ for a unique $n \in N$ and a unique $h \in G_\alpha$.

Proof. By (1.1), we have $G = NG_\alpha$. If $n_1, n_2 \in N$ and $h_1, h_2 \in G_\alpha$ and $n_1 h_1 = n_2 h_2$ then $n_2^{-1} n_1 = h_2 h_1^{-1} \in N \cap G_\alpha$. By regularity, $N \cap G_\alpha = 1$, so

$n_1 = n_2$ and $h_1 = h_2$. //

1.6 Theorem. Let G be a transitive permutation group on Ω and N be a normal subgroup regular as a permutation group on Ω. For any element α of Ω, there exists a bijection $\phi = \phi_\alpha : \Omega \to N$ such that, if $g \in G$ and $g = nh$ where $n \in N$ and $h \in G_\alpha$, then

$$\beta g = (h^{-1}(\beta\phi)n\,h)\,\phi^{-1}.$$

Proof. Suppose we are given $\alpha \in \Omega$. For each $\beta \in \Omega$ there exists $m \in N$ such that $\alpha m = \beta$, since N is transitive on Ω. Since in addition N is regular on Ω, m is unique, so we define $\beta\phi_\alpha = \beta\phi = m$; ϕ is a bijection. Now if $g \in G$ and $g = nh$ where $n \in N$ and $h \in G_\alpha$ then n and h are unique by (1.5). But for any $n \in N$ and $\beta \in \Omega$ we have

$$\begin{aligned}\beta n &= (\alpha(\beta\phi))n \\ &= \alpha((\beta\phi)n) \\ &= ((\beta\phi)n)\,\phi^{-1}.\end{aligned}$$

Further, for any $\gamma \in \Omega$ and $h \in G_\alpha$ we have

$$\begin{aligned}\gamma h &= (\alpha(\gamma\phi))\,h \\ &= (\alpha h(h^{-1}(\gamma\phi)\,h) \\ &= (\alpha(h^{-1}(\gamma(\phi)\,h) \quad \text{since } h \in G_\alpha \\ &= (h^{-1}(\gamma\phi)\,h)\,\phi^{-1}.\end{aligned}$$

Combining these two observations, for $\beta \in \Omega$ and $g = nh \in G$ we have

$$\begin{aligned}\beta g = \beta nh &= (\beta n)\,h \\ &= (((\beta\phi)\,n)\,\phi^{-1})\,h \\ &= (h^{-1}(\beta\phi)\,nh)\,\phi^{-1},\end{aligned}$$

since if $\gamma = ((\beta\phi)\,n)\,\phi^{-1}$ then $\gamma h = h^{-1}(\beta\phi)\,nh$. //

One way of looking at this result is that, via the identification ϕ_α of Ω and N, the action of G on N can be described as right multiplication by elements of N together with conjugation by elements of G_α. Thus the significance of the result lies in the fact that it enables us to identify G_α with a group of group-automorphisms of N. This is especially useful where N is a well-behaved abelian group, such as a finitely generated elementary Abelian p-group. Indeed, the whole purpose of these notes is to use the power of representation theory to analyse the structure of permutation groups with "good" abelian normal subgroups.

A *set of imprimitivity* for a transitive permutation group G on Ω is a subset Δ of Ω such that $|\Delta| > 1$ but $\Delta \neq \Omega$ and for each g in G, $\Delta g \cap \Delta \neq \emptyset$ implies $\Delta g \cap \Delta = \Delta$. It is easy to verify (Exercise!) that this condition is equivalent to: for each g in G, $\Delta g \cap \Delta \neq \emptyset$ implies $\Delta g = \Delta$. We use the term "set of imprimitivity" rather than "block" since we want in due course to employ the latter in the context of block designs, and otherwise confusion at that stage would be unavoidable. A permutation group G on Ω is *primitive* if it is transitive and has no set of imprimitivity.

1.7 Proposition. Let G be a transitive permutation group on Ω, and let N be a non-trivial normal subgroup of G. Then N has an orbit Δ which is not a singleton, and if N is not transitive on Ω then any non-singleton orbit of N is a set of imprimitivity for G.

Proof. Since $N \neq 1$ there is an $x \in \Omega$ such that $\{xn : n \in N\}$ is not a singleton, and it is a suitable Δ. Now suppose Γ is any non-singleton orbit of N and that $n \in N$ and $g \in G$. Then

$$\Gamma gn = \Gamma(gng^{-1})g$$

$$= \Gamma g \quad \text{since } gng^{-1} \in N.$$

Hence Γg is also an orbit of N, and since the orbits of a permutation group are equivalence classes, if $\Gamma g \cap \Gamma \neq \emptyset$ then $\Gamma g \cap \Gamma = \Delta$, as required. //

1.8 Corollary. If G is a primitive permutation group on Ω then any non-trivial normal subgroup of G is transitive on Ω. //

1.9 Theorem. Let G be a primitive permutation group on Ω having a non-trivial abelian normal subgroup A. If α is in Ω then

(a) $G = AG_\alpha = G_\alpha A$ and $G_\alpha \cap A = 1$,

(b) there is a bijection $\phi_\alpha : \Omega \to A$,

(c) A is a minimal non-trivial normal subgroup of G,

(d) no proper subgroup of A is normalised by G_α, and

(e) A is an elementary abelian p-group.

Proof. By (1.8) A is transitive on Ω, and this together with (1.3) implies that A is regular on Ω. Since A is transitive (1.1) implies $G = G_\alpha A = AG_\alpha$, and since A is regular $G_\alpha \cap A = 1$. Suppose $1 \neq B \leq A$ and $B \trianglelefteq G$. As in the case of A, we have $G = G_\alpha B = BG_\alpha$, and if $a \in A$ then $a = bh$ for some $h \in G_\alpha$. But $b \in A$ so $h = b^{-1}a \in G_\alpha \cap A = 1$, hence $a = b \in B$. Thus A is a minimal non-trivial normal subgroup of G, and since $G = G_\alpha A$ this is equivalent to condition (a). Since A is abelian, condition (a) implies condition (d). Finally, since A is abelian and a minimal non-trivial normal subgroup, $A^n = 1$ or A for every $n \in \mathbb{N}$ and $\{x \in A : x^m = 1\}$ is 1 or A for every $m \in \mathbb{N}$; it follows that A is an elementary abelian p-group. //

NILPOTENCY

If abelian groups are the best behaved groups then nilpotent groups certainly come next. The structure of finite nilpotent groups is in our context known and this plays an important role in our final results. We begin with the concept of commutator which "measures" the non-commutativity of two elements in a group.

If x and y are elements of a group G, the commutator of x and y is $[x, y] = x^{-1}y^{-1}xy$. Such an element of G is also called a simple commutator of weight 2. If H and K are non-empty subsets of G then $[H, K] = \langle \{ [h, k] : h \in H, k \in K\}\rangle$, the subgroup generated by the simple commutators of weight 2 of form $[h, k]$ where $h \in H$ and $k \in K$.

Note that $[x, y] = 1$ if and only if x and y commute.

1.10 Lemma. If x, y, z are elements of a group G then

(a) $[x, y]^{-1} = [y, x]$,

(b) $[xy, z] = y^{-1}[x, z]y[y, z]$, and

(c) $[x, yz] = [x, z]z^{-1}[x, y]z$.

Proof. In each case the proof is simply to calculate the two sides. It is well worthwhile getting practice in this type of manipulation, so we prove (b), and leave the other assertion to be verified by the interested reader.

$$\begin{aligned}[xy, z] &= (xy)^{-1}z^{-1}xyz \\ &= y^{-1}x^{-1}z^{-1}xyz,\end{aligned}$$

and

$$\begin{aligned}y^{-1}[x, z]y[y, z] &= y^{-1}x^{-1}z^{-1}xzyy^{-1}z^{-1}yz \\ &= y^{-1}x^{-1}z^{-1}xyz. \quad //\end{aligned}$$

1.11 Lemma. Let x, y be elements of a group G, and let $z = [x, y]$. Then
(a) if $[z, x] = 1$ then $[x^r, y] = z^r$ for $r \in \mathbb{Z}$. If $[z, x] = [y, z] = 1$ then
(b) $[x^r, y^s] = z^{rs}$ for $r, s \in \mathbb{Z}$, and
(c) $(yx)^i = z^{i(i-1)/2}y^i x^i$ for $i \in \mathbb{N}$.

Remark. Once again, the reader not fully familiar with this type of manipulation will find it worthwhile to follow the proofs carefully, though in this case we give the full argument.

Proof. (a) $y^{-1}xy = xz$, so for any given $r \in \mathbb{Z}$ $y^{-1}x^r y = (xz)^r = x^r z^r$ since $[x, z] = 1$. Hence $[x^r, y] = z^r$.
(b) From (a), $[x^r, y] = z^r$, so by (1.10) $[y, x^r] = z^{-r}$. But by (a) $[y^s, x^r] = [y, x^r]^s$, so $[x^r, y^s] = [y^s, x^r]^{-1} = z^{rs}$.
(c) We proceed by induction on $i \in \mathbb{N}$. The result is clearly true for $i = 1$, so suppose $i > 1$ and $(yx)^{i-1} = z^{(i-1)(1-2)/2}y^{i-1}x^{i-1}$. Then

$$\begin{aligned}(yx)^i &= yxz^{(i-1)(i-2)/2}y^{i-1}x^{i-1}\\ &= z^{(i-1)(i-2)/2}yxy^{i-1}x^{i-1}\\ &= z^{(i-1)(i-2)/2}yy^{i-1}x[x, y^{i-1}]x^{i-1}\\ &= z^{(i-1)(i-2)/2}y^i xz^{i-1}x^{i-1}\\ &= z^{i(i-1)/2}y^i x^i . \quad //\end{aligned}$$

1.12 Lemma. Let A be a subgroup of a group B. Then A is normal in B if and only if $[A, B] \subseteq A$.

Proof. If A is normal in B and $a \in A$ and $b \in B$ then $a^{-1}b^{-1}ab = a^{-1}(b^{-1}ab) \in A$ so $[A, B] \subseteq A$. Conversely, if $[A, B] \subseteq A$ and $a \in A$, $b \in B$ then $a^{-1}b^{-1}ab \in A$ so $b^{-1}ab \in A$. Hence A is normal in B. //

1.13 Lemma. Let A and B be subgroups of a group G then $[A, B]$ is normal in $\langle A, B\rangle$.

Proof. If $a, c \in A$ and $b, d \in B$ then by (1.10)

$$c^{-1}[a, b]c = [ac, b][c, b]^{-1} \in [A, B]$$

and

$$d^{-1}[a, b]d = [a, d]^{-1}[a, bd] \in [A, B] ,$$

and the result follows straightforwardly. //

For any group G, we let $Z(G)$ denote the centre of G; that is, $Z(G) = \{x \in G : xg = gx \text{ for all } g \in G\}$. Of course $Z(G)$ is a normal subgroup of G. We can now define two series of subgroups of a group G which will be fundamental to much that follows.

The lower central series of a group G is the descending series of subgroups

$$\gamma_1 = \gamma_1(G) \supseteq \gamma_2 = \gamma_2(G) \supseteq \ldots$$

defined recursively by $\gamma_1(G) = G$ and $\gamma_{n+1}(G) = [\gamma_n(G), G]$.

The <u>upper central series</u> of a group G is the ascending series of subgroups

$$Z_0 = Z_0(G) \subseteq Z_1(G) = Z_1 \subseteq \ldots$$

defined recursively by $Z_0(G) = 1$ and $Z_{n+1}(G)/Z_n(G) = Z(G/Z_n(G))$, where $Z_{n+1}(G)/Z_n(G)$ is embedded as a subgroup of $G/Z_n(G)$ in the obvious, "natural" fashion.

The relationship between the two series of subgroups just defined is clarified by

<u>1.14 Proposition.</u> Let G be a group, then for $n \geq 0$ $\gamma_{n+1}(G) = 1$ if and only if $Z_n(G) = G$.

<u>Proof.</u> Suppose first that $\gamma_{n+1}(G) = 1$. We shall show by induction that $Z_r(G) \geq \gamma_{n+1-r}(G)$. Observe that $Z_0(G) = 1 = \gamma_{n+1-0}(G)$; assume that $Z_r(G) \geq \gamma_{n+1-r}(G)$ and $r \geq 1$.

Now $\gamma_{n+1-r-1}(G) = \gamma_{n-r}(G)$ and by definition $[\gamma_{n-r}(G), G] = \gamma_{n-r+1}(G)$ so $[\gamma_{n-r}(G), G] \leq Z_r(G)$. Hence $\gamma_{n-r}(G) Z_r(G)/Z_r(G) \leq Z(G/Z_r(G))$. Thus $Z_{r+1}(G) \geq \gamma_{n-r}(G)$. But $\gamma_1(G) = G$, so $Z_n(G) = G$.

Conversely, suppose that $Z_n(G) = G$. We show by induction that $Z_{n-r}(G) \geq \gamma_{r+1}(G)$. If $r = 0$ then $Z_n(G) = G = \gamma_1(G)$. So suppose that $Z_{n-r-1}(G) \geq \gamma_r(G)$. Then $[Z_{n-r-1}(G), G] \leq Z_{n-r}(G)$ and so

$$\begin{aligned} Z_{n-r}(G) &\geq [Z_{n-r-1}(G), G] \\ &\geq [\gamma_r(G), G] = \gamma_{r+1}(G), \end{aligned}$$

and the result follows. //

A group G is said to be <u>nilpotent</u> if there exists an integer $c \geq 1$ such that $\gamma_{c+1}(G) = 1$ or equivalently $Z_c(G) = G$. If c is such that $\gamma_c(G) \neq \gamma_{c+1}(G) = 1$, then G is said to have (<u>nilpotency</u>) <u>class</u> c.

<u>1.15 Lemma.</u> Let $G_1, \ldots, G_n$ for $n \geq 1$ a natural number be nilpotent groups of classes $c_1, \ldots, c_n$ respectively. Then their direct product G is a

nilpotent group of class c where $c = \max\{c_1, \dots, c_n\}$.

Proof. Clearly $\gamma_1(G) \simeq \gamma_1(G_1) \times \gamma_r(G_2) \times \dots \times \gamma_1(G_n)$, and it is not difficult to see that

$$\gamma_r(G) \simeq \gamma_r(G_1) \times \gamma_r(G_2) \times \dots \times \gamma_r(G_n),$$

for each integer r. Clearly $\gamma_c(G_i) \neq \gamma_{c+1}(G_i) = 1$ for some i, and $\gamma_{c+1}(G_i) = 1$ for each i, $1 \leq i \leq n$, so $\gamma_c(G) \neq \gamma_{c+1}(G) = 1$. That is, G is nilpotent of class c. //

1.16 Theorem. For a finite group G the following conditions are equivalent:

(i) G is nilpotent;

(ii) if $A \neq G$ is a subgroup of G then $A \neq N_G(A)$;

(iii) each maximal subgroup of G is normal in G;

(iv) G is the direct product of its Sylow subgroups.

Proof. (i) $\Rightarrow$ (ii). Let i be chosen so that $\gamma_{i+1}(G) \leq A$ and $\gamma_i(G) \nleq A$. Then $[A, \gamma_i(G)] \leq [G, \gamma_i(G)] \leq \gamma_{i+1}(G) \leq A$, so $[A, \gamma_i(G)] \leq A$. Hence $[A, A\gamma_i(G)] \leq A$, and so by Lemma 1.12 A is normal in $A\gamma_i(G)$, which strictly contains A.

(ii) $\Rightarrow$ (iii). If M is a maximal subgroup of G then $M \neq N_G(M)$ so $N_G(M) = G$.

(iii) $\Rightarrow$ (iv). Let P be a Sylow subgroup of G. If $N_G(P) \neq G$ let M be a maximal subgroup of G containing $N_G(P)$. Since M is normal in G we have $G = MN_G(P)$ by the Frattini argument. But this is a contradiction since $N_G(P) \leq M$, so P is normal in G. An easy inductive argument then shows that, since every Sylow subgroup is normal, G is the direct product of its Sylow subgroups.

(iv) $\Rightarrow$ (i). If P is a finite p-group for the rational prime p, counting the elements of conjugacy classes shows that $Z(P) \neq 1$. It follows at once that for some integer $t > 1$ $Z_t(P) = P$, so by Proposition 1.14 P is nilpotent. The result then follows from Lemma 1.15. //

__1.17 Theorem (Fitting).__ Let G be a group and, for some natural number $n \geq 1$, $H_1, \ldots, H_n$ be nilpotent normal subgroups of classes $c_1, \ldots, c_n$ respectively. Then $H_1 \ldots H_n$ is a nilpotent normal subgroup of G, of class $c \leq \sum_1^n c_i$.

__Proof.__ We will prove the result for $n = 2$; the general case then follows easily by induction. Let $H_1 = H$ and $H_2 = K$ and their classes be c and d respectively.

If L is a normal subgroup of G then $[HK, L] = [H, L][K, L]$ from the results in Lemma 1.10, so $[HK, HK] = [H, K][H, H][K, K]$. It follows fairly easily that

$$\begin{aligned}\gamma_r(HK) &= [HK, \ldots, HK] \qquad r \text{ terms}\\ &= \Pi[L_1, \ldots, L_r]\end{aligned}$$

where the product is taken over terms which are commutators of r terms and each term L_j is either H or K. Since H and K are normal in G so are $\gamma_r(H)$ and $\gamma_r(K)$. Thus if, in some $[L_1, \ldots, L_r]$, H occurs as a term j times and K occurs $r - j$ times we get

$$[L_1, \ldots, L_r] \leq \gamma_j(H) \cap \gamma_{r-j}(K) .$$

Thus

$$\gamma_{c+d+1}(HK) \leq \Pi\,(\gamma_j(H) \cap \gamma_{c+d+1-j}(K)) = 1. \quad //$$

__1.18 Corollary.__ Let G be a group with at most finitely many normal nilpotent subgroups (for example, let G be finite). Then G has a unique maximal normal nilpotent subgroup.

__Proof.__ If $N_1, \ldots, N_r$ are all the normal nilpotent subgroups of G, let $F = F(G) = \Pi N_i$. By the Theorem F is itself nilpotent and normal, and clearly F contains every nilpotent normal subgroup. //

The unique maximal normal subgroup of this corollary is called the Fitting subgroup of G and denoted by $F(G)$.

A group G is said to be soluble (in American usage, solvable) if there exists a finite set of subgroups $H_0, \dots, H_n$ such that

$$1 = H_0 \leq H_1 \leq \dots \leq H_n = G$$

and, for each $i < n$, H_i is normal in H_{i+1}, and also the quotient group H_{i+1}/H_i is abelian. A group G is said to be simple if $G \neq 1$ and G has no non-trivial normal subgroups, or equivalently G has no non-trivial quotient groups.

It is straightforward to check that every nilpotent group is soluble, though the reverse statement is not true. There are groups which are not soluble, for example any non-abelian simple group.

A set $H_0, \dots, H_n$ of subgroups of a group G is a chief series if

$$1 = H_0 \leq H_1 \leq \dots \leq H_n = G$$

and, for each $i < n$, H_i is normal in G and $H_i \neq H_{i+1}$ and if $H_i \leq K \leq H_{i+1}$ and K is normal in G then $H_i = K$ or $K = H_{i+1}$. The quotient groups H_{i+1}/H_i are called chief factors of G, and it can be shown that if G has a chief series then any two such series have the same number of terms and (up to order) the same chief factors. There are, of course, groups which do not have chief series; for example, the infinite cyclic group does not. It is clear that every finite group has a chief series, since we can pick successive subgroups such that each group after 1 is minimal among the normal subgroups that strictly contain the previous group. Indeed a group is easily seen to have a chief series if and only if it satisfies the ascending and descending chain conditions on normal subgroups. A closely analogous result for modules will be proved in Chapter 2.

1.19 Theorem. Let G be a finite soluble group. Then $Z(F(G)) = C_G(F(G)) \trianglelefteq F(G)$.

Proof. Let $K = F(G)$ and $H = C_G(K)$, and suppose $H \not\leq K$. Then $H \cap K = Z(F(G))$ and we can find a subgroup J of G such that $H \cap K \lhd J \trianglelefteq H$ and $J/(H \cap K)$ is a chief factor of G. Since normal subgroups of a soluble group are easily seen to be soluble $\gamma_2(J) \leq H \cap K$. But $H \cap K = Z(F(G)) \leq Z(C_G(F(G)))$, so $\gamma_3(J) = [\gamma_2(J), J] \leq [Z(F(G)), J] = 1$. Hence J is a normal nilpotent subgroup of G not contained in $F(G)$. This contradiction proves the theorem. //

A subgroup F is said to be a characteristic subgroup of a group G if $Fs \subseteq F$ for every automorphism s of G. For example, the centre $Z(G)$ of a group G is a characteristic subgroup.

1.20 Lemma. If G is a finite group and H is a normal subgroup of G then $F(H) \leq F(G)$.

Proof. It is clear that the Fitting subgroup of a group (when it exists) is a characteristic subgroup. Hence $F(H)$ is normal in G, so $F(H) \leq F(G)$. //

1.21 Corollary. Let G be a finite group. Then $F(G)\,C_G(F(G))/F(G)$ has no non-trivial normal soluble subgroup.

Proof. Let $A/F(G)$ be a normal soluble subgroup of $F(G)\,C_G(F(G))/F(G)$. Then A is soluble and (since A is normal in G) $F(A) = F(G)$. But by Theorem 1.19 $C_A(F(A)) \leq A$. Since $A \leq C_G(F(G))\,F(G)$ and $F(G) \leq A$ we have $A = F(G)$. //

Finally in this section we introduce one more subgroup, the Frattini subgroup of a group G. We begin by defining a non-generator element x of a group G: $x \in G$ is said to be a non-generator if, whenever X is a subset of G and $G = \langle X \cup \{x\} \rangle$ then $G = \langle X \rangle$. There is a close analogy between this definition and the definitions (see Chapter 2) of a superfluous element and a superfluous submodule(t) of a module(t), though these latter concepts are of much more recent date than Frattini's.

1.22 Lemma. Let G be a group and let $\Phi(G)$ denote the set of non-generators of G. Then $\Phi(G)$ is a characteristic subgroup of G.

Proof. If x is a non-generator then so is x^{-1}. If x and y are non-generators and $G = \langle X \cup \{xy\}\rangle$ then $G = \langle X \cup \{xy\} \cup \{x\}\rangle = \langle X \cup \{y\} \cup \{x\}\rangle = \langle X \cup \{y\}\rangle = \langle X\rangle$, so $xy \in \Phi(G)$. Clearly the property of being a non-generator is invariant under the action of any automorphism. //

The subgroup characterised in Lemma 1.22 is called the Frattini subgroup of G, and denoted by $\Phi(G)$. The following result shows that the Frattini subgroup has, for certain groups, an alternative characterisation which closely resembles that of the Jacobson radical of a ring(let) which we shall meet in Chapter 2.

1.23 Proposition. Let G be a group. If G has at least one maximal subgroup then $\Phi(G) = \cap M$, where M runs over all the maximal subgroups of G. Otherwise G has no maximal subgroups, and $\Phi(G) = G$. If G has a.c.c. on subgroups, for example if G is finite, then $\Phi(G) = \bigcap_{M \text{ a maximal subgroup}} M \neq G$.

Proof. Suppose G has a maximal subgroup. Let W denote the intersection of all the maximal subgroups of G. Let $x \in W$ and let $G = \langle X \cup \{x\}\rangle$, where $G \neq \langle X\rangle$. If M is a maximal subgroup containing X, then $x \in M$ so $G = \langle X \cup \{x\}\rangle \leq M$; contradiction. Hence $G = \langle X\rangle$; that is x is a non-generator. Conversely, suppose $y \in \Phi(G)$, and N is a maximal subgroup of G. If $y \notin N$ then $G = \langle N \cup \{y\}\rangle \neq \langle N\rangle$; contradiction. Hence $y \in W$, so finally $W = \Phi(G)$ whenever G has at least one maximal subgroup.

If G has no maximal subgroups, suppose $x \in G$ and $G = \langle X \cup \{x\}\rangle$ for some $X \subseteq G$. If $G \neq \langle X\rangle$ then by Zorn's Lemma we can find a subgroup L of G such that L is maximal with respect to $X \subseteq L$ but $x \notin L$. But then if $L \subseteq K \subseteq G$ for a subgroup K of G then $L = K$ or $K = G$, so L is a maximal subgroup of G; contradiction.

Hence $\Phi(G) = G$. //

There are groups which do not have any maximal subgroups. The additive group $\mathbb{Q}^+$ has no maximal subgroups, and every element is a non-generator, so $\Phi(\mathbb{Q}^+) = \mathbb{Q}^+$.

The following result connects the Frattini subgroup and nilpotence in the case of finite groups.

1.24 Proposition. If G is a finite group then $\Phi(G)$ is nilpotent.

Proof. Let P be a Sylow p-subgroup of $\Phi(G)$ for some prime p. Then $G = \Phi(G)N_G(P)$ by the Frattini argument. But since $\Phi(G)$ consists of non-generators, $G = N_G(P)$. So each Sylow subgroup of $\Phi(G)$ is normal in G and hence $\Phi(G)$ is nilpotent by Theorem 1.16. //

AUTOMORPHISMS

Suppose that H and K are groups and s is a homomorphism from K to $\mathrm{Aut}(H)$, the group of automorphisms of H. Then we can construct a new group G, the split extension (or semidirect product) of H by K, as follows.

Let $G = \{(h, k) : h \in H, k \in K\}$ and define multiplication in G by

$$(h, k)(h_1, k_1) = (h(h_1(k^{-1}s)), kk_1)$$

for each pair (h, k), (h_1, k_1) of elements of G. It is routine to check that with this definition of product G is a group. It has two particular subgroups $H' = \{(h, 1) : h \in H\}$ and $K' = \{(1, k) : k \in K\}$ and again it is routine to show that $H' \simeq H$, $K' \simeq K$ (with the obvious isomorphisms), that H' is normal in G and that $G = H'K'$ with $H' \cap K' = 1$.

Any group P which can be written $P = LM$ for subgroups L and M such that L is normal in P and $L \cap M = 1$ is said to be a split extension of L by M. It is straightforward to check that, in this situation, if $s : M \to \mathrm{Aut}(L)$ is the homomorphism which takes $m \in M$ to the L-automorphism "conjugation by m", then P is (isomorphic to) that split extension of L by M which is defined using s.

The split extension is a very useful way of constructing groups as well as being helpful in studying automorphism groups. Given a group H and a subgroup K of Aut(H) then, using the inclusion homomorphism we can construct a split extension HK of H by K such that H and K are "naturally" embedded in HK.

A torsion group is a group in which every element has finite order. Given a non-empty set Π of (positive rational) primes a group G is called a Π-group if every element of G other than 1 has order divisible only by primes in Π. Obviously a Π-group is a torsion group. We shall use the customary convention that Π' denotes the complementary set of positive rational primes. If $\Pi = \{ p \}$ we put $\Pi' = p$, rather than $\{p\}$. If a is an automorphism of a group H and $h \in H$ we shall frequently denote the image of h under a by h^a (this is the so-called "exponential" notation). Given a group G and a subgroup A of the automorphism group of G we write $[G, A] = \langle \{ g^{-1}.g^a : g \in G, a \in A \}\rangle$. If we embed G and A in the split extension H = GA of G by A, then $[G, A]$ is precisely the subgroup generated by $\{ g^{-1}a^{-1}ga : g \in G, a \in A \}$, and indeed $[G, A]$ is the subgroup generated by commutators of H, in the sense discussed at the start of the section on nilpotency.

A subgroup X of G is called A-invariant if it satisfies any of the three equivalent conditions: $Xa = X$ for each $a \in A$, $Xa \subseteq X$ for each $a \in A$, $X \subseteq Xa$ for each $a \in A$. If X is A-invariant then the restriction map induces a homomorphism of A to Aut(X). Note that $[G, A]$ is A-invariant. This terminology and notation is used in the following result.

1.25 Lemma. Let G be a Π-group and A be a torsion subgroup of the automorphism group of G. Suppose that G has a series of A-invariant subgroups,

$$G = G_0 \geq G_1 \geq \dots \geq G_n = 1$$

for some natural number n, such that $[G_i, A] \leq G_{i+1}$ for $0 \leq i < n$. Then A is a Π-group.

Proof. We regard all the groups as being embedded in the split extension GA of G by A, and then proceed by induction on n. If $n = 1$ then $ag = ga$ in GA for each $g \in G$ and $a \in A$, so $A = 1$. So assume that the result is true for $n - 1 \geq 0$, and let $b \in A$ not be a Π-element. If $b \neq 1$ then the cyclic group $\langle b \rangle$ contains a Π'-element $a \neq 1$ (since A is a torsion group). By the inductive hypothesis a acts trivially on G_1, so $[g, a] = 1$ for all $g \in G_1$. Let $g \in G$. Then $[g, a] = g_1 \in G_1$. So since $[g_1, a] = 1$ we have $[g, a^r] = g_1{}^r$ by Lemma 1.11 so the order of g_1 divides the order of a. Since G is a Π-group this implies $g_1 = 1$. But that in turn implies $g^a = g$ so that a is the identity map; contradiction. It follows that A is a Π-group. //

If, with the same notation, we consider the extension GA of G by A then $[G, A]$ is normal in GA by Lemma 1.13. If A is a Π'-group when G is a Π-group then Lemma 1.25 has an interesting consequence.

1.26 Lemma. If G is a Π-group and A is a Π'-subgroup of $\text{Aut}(G)$ then $[G, A, A] = 1$ implies $A = 1$.

Proof. Clearly the series

$$G \geq [G, A] \geq [G, A, A] = 1$$

satisfies the hypotheses of Lemma 1.25, so A is both a Π- and a Π'-group, i.e. $A = 1$. //

1.27 Theorem. Let G be a finitely generated group and assume $\Phi(G)$ is a finite Π-group. Let B be the subgroup of $\text{Aut}(G)$ whose elements induce the identity map on $G/\Phi(G)$. Let d be the least integer such that G can be generated by d elements. Then B is finite and $|B| \,\Big|\, |\Phi(G)|^d$, so B is a Π-group.

Proof. Let $g_1, \ldots, g_h$ be elements of G. Then from the definition of $\Phi(G)$ it is clear that $g_1, \ldots, g_h$ generate G if and only if $\bar{g}_1, \ldots, \bar{g}_h$ generate $\bar{G} = G/\Phi(G)$, where for $1 \leq i \leq h$, $\bar{g}_i = g_i\Phi(G)$. So suppose $g_1, \ldots, g_d$ is a generating set for G. Let Ω be the collection $\{\{g_1u_1, \ldots, g_du_d : u_i \in \Phi(G)$ for

$1 \le i \le d\}\}$. Clearly Ω is a collection of generating sets of G that contain d elements, and equally clearly $|\Omega| = |\Phi(G)|^d$.

Since B acts on Ω (because B acts trivially on $G/\phi(G)$) the size of the orbit of $\omega \in \Omega$ under B is $[B : B_\omega]$, which is thus finite. But if $\omega = \{z_1, \ldots, z_d\}$ and $b \in B_\omega$ then $z_i^b = z_i$ for $1 \le i \le d$, since b induces the identity on $G/\Phi(G)$. Hence since $z_1, \ldots, z_d$ is a generating set we have $b = 1$. Hence $B_\omega = 1$, so each orbit in Ω is of order $|B|$, which is finite. Thus $|B| \mid |\Phi(G)|^d$. //

1.28 Corollary. If P is a finite p-group and A is a non-trivial p'-group of automorphisms of P, then $[P, A] \not\leq \Phi(P)$. //

If B is an abelian group, it is frequently written additively. An endomorphism of a group G is a homomorphism of G to itself. If B is an (additive) abelian group and $E = \mathrm{End}(B)$ is the set of endomorphisms of B, we can define an addition on E by

$$b(f + g) = (bf) + (bg) \qquad \text{for } b \in B,\ f,\ g \in E$$

and take as multiplication the ordinary composition of functions. With these conventions $E = \mathrm{End}(B)$ is an associative ring with unity, the endomorphism ring of B. An element e of a ring R is idempotent if $e^2 = e$. (See Chapter 2 for further discussion of rings.)

1.29 Lemma. Let B be an additive abelian group and e be an idempotent of $E = \mathrm{End}(B)$. Then $B = Be \oplus B(1 - e)$, where $Be = \{be : b \in B\}$ and

$$\begin{aligned} B(1 - e) &= \{b - be : b \in B\} \\ &= \{b(1 - e) : b \in B\}. \end{aligned}$$

Note. The symbol $\oplus$ here denotes direct sum, the analogue in additive notation of the direct product of two groups that are written multiplicatively.

Proof. If $b \in B$ then $b = be + b(1 - e)$, so $B = Be + B(1-e) = \{be + c(1-e) : b, c \in B\}$. If $x \in Be \cap B(1 - e)$ then $x = x'e$ for some x', so $x = x'e = x'e^2 = xe$, and $x = x''(1 - e)$ for some x'', so $xe = x''(1 - e)e = x''(e - e^2) = x''(e - e) = 0$. Hence $Be \cap B(1 - e) = 0$. //

If G is a group and A is a group of automorphisms of G then $C_G(A) = \{x \in G : x^a = x \text{ for all } a \in A\}$. It is the centraliser in G of A if we form the split extension GA of G by A.

1.30 Theorem. If B is a finite additive abelian Π-group and A is a finite Π'-group of automorphisms of B then

$$B = [B, A] \oplus C_B(A) .$$

Proof. Let $n = |A|$. Since B is a Π-group and A is a finite Π'-group the map $\hat{n} : B \to B$ defined by $x\hat{n} = xn$ for $x \in B$ is an injective endomorphism of B. Since B is finite, it has an inverse, which we can denote by $\frac{1}{n}$.

Now let $e = \frac{1}{n}(\sum_{a \in A} a) \in E = \mathrm{End}(B)$. Clearly $ea = ae = e$ for any $a \in A$, so $e^2 = \frac{1}{n}\sum_a ea = \frac{1}{n}\Sigma e = e$. By Lemma 1.29 $B = Be \oplus B(1 - e)$. Now if $b \in Be$ then $b = ce$ for some $c \in B$, and then $ba = cea = ce = b$. Conversely, if $ba = b$ for all $a \in A$ then $be = b$. Thus $Be = C_B(A)$.

On the other hand if $b \in B(1 - e)$ then $b = d(1 - e)$ for some $d \in B$. Now

$$1 - e = \hat{n} . \frac{1}{n} - e = \hat{n} . \frac{1}{n} - \frac{1}{n} \sum_{a \in A} a$$

$$= \frac{1}{n}(\sum_{a \in A} (1 - a)) .$$

But $d(1 - a) = d - da$

$$= [-d, a] \in [B, A]$$

so $b = \frac{1}{n}(\sum_{a \in A} d(1 - a) \in [B, A]$ so $B(1 - e) \leq [B, A]$. Conversely, if $b \in B$ and $a \in A$ then $[b, a] = b(-1 + a)$ so $[b, a]e = b(-1 + a)e = b(-e + e) = 0$. But $[b, a] = [b, a]e + [b, a](1 - e)$ so $[b, a] \in B(1 - e) = \{x \in B : xe = 0\}$, and the result is proved. //

EXTRASPECIAL p-GROUPS

A finite p-group G is called special either if it is an elementary abelian p-group or if $\Phi(G) = G' = Z(G)$ and $Z(G)$ is elementary abelian. A finite p-group G is called extraspecial if it is a non-abelian special group and $|Z(G)| = p$.

The reasons for these definitions go back to the realisation that the groups involved are in a certain sense "minimal". Extraspecial groups are the "smallest" non-abelian p-groups that one can have. In many inductive situations it is possible to reduce the problem to one which involves considering extraspecial groups.

It is also worth pointing out that for an extraspecial group G the quotient group $G/Z(G)$ can be thought of as a vector space over the field $GF(p)$ of p elements. For elements x, y of $G/Z(G)$ let x', y' be any elements of G which map to x, y respectively under the natural map. Let z be a generator of $Z(G)$. Then the mapping $(x, y) \to r$ where $[x', y'] = z^r$ from $G/Z(G) \times G/Z(G)$ to $GF(p)$ is a form on $G/Z(G)$, with the properties (deliberately mixing additive and multiplicative notations) that

$$(x, y) = -(y, x)$$

and $$(xy, z) = (x, z) + (y, z)$$

$$(x, yz) = (x, y) + (x, z)$$

for any x, y, z in $G/Z(G)$. This observation plays an important role in the structure of the extraspecial groups.

The following important technical lemma is due to Hering [13].

1.31 Lemma. If G is a finite p-group such that every proper characteristic subgroup of G is cyclic and contained in $Z(G)$ then one of the following statements holds:

(a) G is cyclic;

(b) G is elementary abelian,

(c) G is extraspecial;

(d) $p = 2$ and $G = TZ(G)$ where T is an extraspecial 2-group of order ≥ 16,

$Z(G)$ is cyclic of order 4 and $T \cap Z(G) = Z(T)$ is of order 2.

Proof. Suppose G is abelian but not elementary abelian. The maximal elementary abelian subgroup of G is characteristic, hence cyclic, and so G is cyclic.

So suppose that G is non-abelian. Clearly $G' \leq \Phi(G) \leq Z(G)$ and $Z(G)$ is cyclic. Hence $[a, b]^p = [a, b^p] = 1$ for all $a, b \in G$. So G' is elementary abelian, and then $1 \neq |G'| = p$ since $Z(G)$ is cyclic and $1 \neq Z(G) \neq G$. If $G' \neq \Phi(G)$ then $\{x \in G : x^p \in G'\}$ is a proper characteristic subgroup of G, and therefore cyclic; in consequence, G/G' is cyclic, and thus G is cyclic; contradiction.

Hence if G is non-abelian $G' = \Phi(G)$. Also, if $G' \leq Z(G)$ then by Lemma 1.11 $(xy)^i = x^i y^i [x, y]^{i(i-1)/2}$ for $x, y \in G$ and $i \geq 0$.

If $p > 2$, let B denote the set of elements $g \in G$ such that $g^p = 1$. Since, for $x, y \in G$, $(xy)^p = x^p y^p [x, y]^{p(p-1)/2}$ and $|G'| = p$, B is a characteristic subgroup of G. If $B \neq G$ the B and hence G are cyclic; contradiction. So $B = G$, so $Z(G) = G'$. It follows that G is extraspecial.

If $p = 2$, let C denote the set of elements $h \in G$ such that $h^4 = 1$. By a similar argument C is a characteristic subgroup of G, so $C = G$ as above. Hence either $Z(G) = G'$ and G is extraspecial or $|Z(G)| = 4$.

If $|Z(G)| = 4$ let T be a subgroup such that T/G' is a complement of $Z(G)/G'$ in the elementary abelian group G/G'. Then clearly $G = T.Z(G)$ and $[T, Z(G)] = 1$ and $T \cap Z(G) = G'$. In addition $G = T.Z(G)$ implies $Z(T) \leq Z(G)$. Since $C = G$ there exist elements t_1 and t_2 of G such that $|\langle t_1 \rangle| = |\langle t_2 \rangle| = 4$ and $G = \langle t_1 \rangle \langle t_2 \rangle Z(G)$.

But $\langle t_1 \rangle \langle t_2 \rangle / G'$ is a complement of $Z(G)/G'$, so from above it is extraspecial. Thus $\langle t_1 \rangle \langle t_2 \rangle$ is a quaternian group, and it is characteristic in G since $C \subseteq \langle t_1 \rangle \langle t_2 \rangle \cup Z(G)$. Consequently $|T| > 8$. //

2 Basic ring theory

INTRODUCTION

In this chapter we shall render unambiguous our usage of various ring- and module-theoretic terms; we should apologise for the time spent on doing this but it is hoped that the effort is worthwhile. We prove a number of important results concerning Jacobson radicals, the descending chain condition on one sided ideals and semisimple modules. A number of techniques are also developed which are important in later chapters.

The flavour of our treatment is "modern" in a fairly precise sense: in particular we try to avoid chain conditions through making the Jacobson radical (in a somewhat unorthodox treatment) central to our approach. In addition we use techniques (such as concentrating on direct sums of subobjects) associated with the name of A. W. Goldie [9] in our proofs of such classical results as the Artin-Wedderburn theorem (on simple algebras with right d. c. c.) which we develop in an unfamiliar but clear and economical fashion.

Whilst we have tried to write the proofs in this chapter with the wider considerations of ring theory taken seriously this is not true of the theorems presented. These have been chosen quite narrowly for their application in the group theory and representation theory needed for the later theorems we seek to prove in this book.

DEFINITIONS AND CONVENTIONS

In modern algebra there are three key concepts: group, category, and "module over a ring". Each of the first two terms has been precisely defined, in general mathematical parlance, but the terms "ring" and "module" continue - remarkably - to lack almost universally accepted definitions. The two sources of problems are associativity and the existence (and behaviour) of a unity element (if any).

Throughout this book (including the brief mention in Chapter 1), the term <u>ring</u> means "associative ring with unity". On the occasions on which we want to refer to an "associative ring not necessarily with unity", we shall speak of a <u>ringlet</u>. Similarly, a <u>module</u> will always be a (left or right) unital module over a ring (which will usually be specified, or clear from the context). Let R and S be rings then M is an <u>R-S bimodule</u> if M is endowed with a left structure as R-module and a right structure as S-module in an <u>associative</u> fashion; that is

$$(rm)s = r(ms) \ \forall r \in R \ \forall m \in M \ \forall s \in S.$$

Once again, on rare occasions we shall want to refer to a <u>modulet</u>, or a <u>bi-modulet</u>, by which terms we denote a "not necessarily unital" module or bimodule over a ringlet R or a pair R, S of ringlets. (We shall, on the other hand, never vary the assumption that two-sided objects are associative.)

Whilst appreciating that this involves yet more terminology, we feel that the gain in clarity is worth the sacrifice involved. Readers familiar with the concept of a category will be able to check that, if R is a ring, the categories of (right) modules and of (right) modulets can have drastically different properties. For example, if R is a field of characteristic p then every right R-module M has an elementary abelian p-group as its underlying additive group. By contrast, the underlying additive group of a right R-modulet can be a torsion free divisible abelian group.

If R is a ringlet, the <u>centre</u> Z(R) of R is the set $\{x \in R : xa = ax \ \forall a \in R\}$. It is a ringlet contained in R, and if R is a ring then Z(R) is a <u>subring</u> in the sense (which we shall always use) that T is a subring of R if and only if T, R are rings, $T \subseteq R$, and T and R have the same unity element. By contrast, if U, S are ringlets, and $U \subseteq S$ then U is a <u>subringlet</u> of S, whether or not the unity element(s), if any, of U and S coincide. If R and S are rings and if R is included in S as a direct summand, and if $R \neq S$ then R is a subringlet of S but not a subring.

To avoid endless exceptions, we shall assume that, unless otherwise stated, a ring R is non-zero, in which case the unity element $1 = 1_R$ is itself non-zero.

A ringlet, on the other hand, may be zero. The term ideal always means a two-sided ideal.

A ring or ringlet is prime if it is non-zero and the product of two non-zero ideals is a non-zero ideal. It is a domain if it is non-zero and the product of two non-zero elements is a non-zero element. Thus a domain is prime, but a prime ring need not be a domain. A ring or ringlet R is of course commutative if $R = Z(R)$, and for commutative ringlets the concepts of prime ringlet and of domain coincide.

If $R \subseteq S$ are rings we say that $R \subseteq S$ is a ring extension. If C is a commutative ring and $f : C \to S$ is a homomorphism of C into a ring S, then S is a C-algebra if $Cf \subseteq Z(S)$. Conversely, if we refer to a ring R as a C-algebra we shall take that to mean that C is commutative and that for a particular (though possibly unspecified) homomorphism $f : C \to R$ $Cf \subseteq Z(R)$. For any embedding g of the complex numbers $\mathbb{C}$ in the real quaternions $\mathbb{H}$, $\mathbb{C}g \not\subseteq Z(\mathbb{H})$, since the image of i is non-central, so there are homomorphisms of commutative rings into non-commutative rings which do not give rise to algebras.

If R is a ring, a unit of R is an element $x \in R$ such that, for some $y \in R$, $xy = yx = 1_R$. The element y is called the inverse of x, since it is easily seen to be unique. The set of units of a ring R (which always contains 1_R) is a multiplicative group, called the group of units of R.

A division ring is a ring in which every non-zero element has a two-sided inverse. The real quaternions are a division ring.

Let R be a ringlet and S be a non-empty subset of a left R-modulet M. Then $LA_R(S) = LA(S) = \{x \in R : xs = 0 \ \forall s \in S\}$ is the annihilator of S. It is a left ideal of R. If $S = M$ then $LA_R(S)$ is a two-sided ideal of M, since $xM = 0$ implies $(xR)M = 0$. If $S = \{m\}$ for $m \in M$ we write $LA(m)$ not $LA(\{m\})$. The right analogues of these objects are denoted by $RA(S)$, $RA(m)$ respectively.

RINGS AND MODULES OF FRACTIONS

Let C be a commutative ring with unity 1. A subset S of C with $0 \notin S$ is multiplicatively closed if $1 \in S$ and $x, y \in S$ implies $xy \in S$. An element x of a ring R is nilpotent if there exists $n \geq 1$ such that $x^n = 0$, and if y is a non-nilpotent element of a commutative ring C then the set $S = \{1, y, y^2, \ldots\} = \{y^r\}_0^\infty$ is a multiplicatively closed set in C. Let D be a commutative integral domain with unity then $T = \{x \in D : x \neq 0\}$ is a multiplicatively closed set. The construction of the ring of fractions of D with respect to T is familiar: it gives rise to the field of fractions K of D in the usual way.

For general commutative rings we can define a similar ring of fractions of a commutative ring C with respect to a multiplicatively closed set S. Thus we consider the cartesian product $C \times S$, and define an equivalence relation $\sim$ on $C \times S$ by

$$\forall c, d \in C \; \forall s, t \in S, \quad (c, s) \sim (d, t)$$

if and only if there exists $u \in S$ such that $(ct - sd)u = 0$.

It is easy to see that $\sim$ is reflexive and symmetric. If $(c, s) \sim (d, t)$ and $(d, t) \sim (e, r)$, in the obvious notation, then for some u and v in S we have

$$(dr - et)v = 0 = (ct - sd)u .$$

Hence $(dr - et)suv = 0 = (ct - sd)ruv$, so $ctruv = etsuv$, so $(cr - es)tuv = 0$; that is, $(c, s) \sim (e, r)$. Thus $\sim$ is also transitive, and hence an equivalence relation.

We denote by $S^{-1}C$ the set of equivalence classes under $\sim$, and let c/s stand for the equivalence class of (c, s), where $c \in C$ and $s \in S$. It is straightforward to check that if we define an addition and a multiplication on $S^{-1}C$ by (respectively, in the obvious notation)

$$c/s + d/t = (ct + ds)/(st)$$

and

$$c/s \times d/t = (cd)/(st),$$

then these operations are well-defined (independently of the choices of representatives of the equivalence classes concerned), and that under these operations $S^{-1}C$ is a ring, with unity element $1/1$.

For example, to verify that multiplication is well-defined, it clearly suffices to consider one of the equivalence classes on the left of the definition, so suppose $(c/s) = (k/r)$ in the obvious notation. Then $(cr - ks)u = 0$ for some $u \in S$, so

$$(cr - ks)udt = (cd.rt - kd.st)u = 0,$$

that is $(cd)/(st) = (ck/rt)$ as required. The reader unfamiliar with this construction will find it useful to verify other details. One additional property worth mention is that the map $f : C \to S^{-1}C$ which takes $a \in C$ to $a/1$ is a ring homomorphism. In fact f has the property that, for each $s \in S$, sf is a unit of $S^{-1}C$. Furthermore, if $g : C \to T$ is any ring homomorphism such that sg is a unit for all $s \in S$ then there exists a unique ring homomorphism $g^* : S^{-1}C \to T$ such that $ag = afg^*$ for every $a \in C$. Since any such pair (g, T) can be "factored through" $(f, S^{-1}C)$, we can in that sense think of the latter as "universal" among ring homomorphisms which take S into the group of units of another ring.

Let R be a ring and M an $R - R$ bimodule, then an element $x \in M$ is <u>centralising</u> if $rx = xr$ for each $r \in R$. The unity and zero elements of a ring R are centralising in the bimodule ${}_RR_R$; in fact the centralising elements of ${}_RR_R$ are simply the elements of $Z(R)$. When R is a commutative ring, any one-sided R-module M may be converted into an $R - R$ bimodule in which multiplication on the other side is simply defined by

$$rm = mr \ \forall r \in R \ \forall m \in M,$$

i.e. by requiring every element to be centralising. Thus, for a commutative ring R and a module M it is possible without difficulty to regard M as a right or a left R-module, or as an $R - R$ bimodule in which every element is centralising. We shall systematically do so without further detailed discussion.

Let C be a commutative ring and M a (left) C module and $S \subseteq C$ be a multiplicatively closed set. We can define an equivalence relation $\sim$ on the cartesian product $M \times S$ in the same way as in the construction of $S^{-1}C$. If $S^{-1}M$ denotes the set of equivalence classes then we can obtain, just as in the ring case, well-defined operations of addition in $S^{-1}M$ and scalar multiplication by elements of $S^{-1}C$. Thus we obtain a left module over the commutative ring $S^{-1}C$, which can also be thought of as a bimodule in which every element is centralising (over $S^{-1}C$).

In particular, when A is a C-algebra over a central subring C and $S \subseteq C$ is multiplicatively closed, we can define (in the obvious way) multiplication in $S^{-1}A$, as well as addition and scalar multiplication. Then the map $(c/s) \to (c/s)(1_A/1_A)$ is an injective ring homomorphism of $S^{-1}C$ into the centre of $S^{-1}A$, so $S^{-1}A$ is an $S^{-1}C$ algebra. We identify $S^{-1}C$ with its image under this map.

<u>2.1 Lemma.</u> Let A be a C-algebra where $C \subseteq Z(A)$ and $S \subseteq C$ be a multiplicatively closed set. For an ideal I of A let

$$I^e = \{x/y \in S^{-1}A : x \in I,\ y \in S\},$$

and for an ideal J of $S^{-1}A$ let $J^c = \{x \in A : x/1 \in J\}$. Then I^e and J^c are ideals of $S^{-1}A$ and A respectively. Also $I \subseteq (I^e)^c$ and $J \subseteq (J^c)^c$.

<u>Proof.</u> If $x_1, x_2 \in I$, $y_1, y_2, s \in S$ and $a \in A$ then

$$\begin{aligned} x_1/y_1 + x_2/y_2 &= (x_1y_2 + y_1x_2)/y_1y_2 \in I^e \\ a/s \,.\, x_1/y_1 &= (ax_1)/sy_1 \in I^e \\ x_1/y_1 \,.\, a/s &= (x_1a)/y_1s \in I^e . \end{aligned}$$

Similarly, if $x, y \in J^c$ and $b \in A$

then $(x + y)/1 = x/1 + y/1 \in J$

and $(bx)/1 = b/1 \,.\, x/1 \in J$

and $(xb)/1 = x/1 \,.\, b/1 \in J$.
The last statements follow trivially. //

It is worth noting that the inclusion $I \subseteq (I^e)^c$ of the previous Lemma can be strict. For example, let $C = A = \mathbb{Z}$, the ring of rational integers, and let $S = \mathbb{Z} - 0$, so that $S^{-1}\mathbb{Z} = \mathbb{Q}$, the rational field. Suppose $I = m\mathbb{Z}$ for some $m \neq 0, \pm 1$. Then $I^e = \mathbb{Q}$, so $(I^e)^c = \mathbb{Z} \neq I$. We do, however, have the following result.

2.2 Proposition. With the notation of the previous Lemma, $(J^c)^e = J$ for every ideal J of $S^{-1}A$.

Proof. We already know $(J^c)^e \subseteq J$. On the other hand, if $x \in A$, $s \in S$ and $x/s \in J$ then $s/1 \,.\, x/s = sx/s = x/1 \in J$, so $x \in J^c$, so $x/s \in (J^c)^e$. //

An ideal P of a ring or ringlet R is said to be prime if the quotient ring R/P is prime. It is a straightforward (and worthwhile) calculation to check that an equivalent definition is: $R - P$ is non-empty and if $x, y \in R - P$ there is an $s \in R$ such that $xsy \in R - P$. The following result shows that prime ideals behave well under the process of forming rings and algebras of fractions.

2.3 Proposition. With the notation of Lemma 2.1, J is a prime ideal of $S^{-1}A$ if and only if J^c is a prime ideal of A such that $J^c \cap S = \emptyset$.

Proof. Suppose J^c is a prime ideal of A and $J^c \cap S = \emptyset$. Then $1 \notin J^c$ so $J \neq S^{-1}A$. Suppose x/s and y/t are elements of $S^{-1}A - J$. Then certainly $x/1$ and $y/1$ are elements of $S^{-1}A - J$. But then $x, y \notin J^c$ so there exists $z \in A$ such that $xzy \notin J^c$. It follows at once that $x/1 \,.\, z/1 \,.\, y/1 \notin J$, and hence that $x/s \,.\, z/1 \,.\, y/t \notin J$, so J is prime. The reverse verification is left to the reader. //

A ring R is said to be local if it has a unique maximal right (or left) ideal $\underline{m}$. (Beware: this is the most common usage, but there are others.)

Let R be local,

and consider the right R-module $R/\underline{m}$. Let $\bar{x} \in R/\underline{m}$ with $\bar{x} \neq 0$. Then as $\underline{m}$ is a maximal right ideal $\bar{x}R = R/\underline{m}$. Let $I = RA_R(\bar{x})$. Let f be the R-module homomorphism of $R \to R/\underline{m}$ given by $r \to \bar{x}r$. The kernel of f is I and so $R/I \cong R/\underline{m}$. Thus I is a maximal right ideal of R. So as $\underline{m}$ is the unique such ideal $I = \underline{m}$. Now if $x \in R$ and $x \in \underline{m}$, $x\underline{m} \subseteq \underline{m}$. If $x \notin \underline{m}$ then $\{y \mid xy \in \underline{m}\}$ is just the annihilator of $x + \underline{m} \in R/\underline{m}$. So $\{y \mid xy \in \underline{m}\} = \underline{m}$ and so $x\underline{m} \subseteq \underline{m}$. Thus $\underline{m}$ is a two-sided ideal of R. So $\underline{m}$ is a two-sided ideal such that $R/\underline{m}$ is a division ring. It is straightforward to check that the following result provides a range of alternative characterisations of a local ring, which we shall use in the sequel.

2.4 Proposition. The following conditions on a ring R are equivalent:

(i) R is local;

(ii) if $r \in R$ either r or $1 - r$ is a unit of R;

(iii) if $r, s \in R$ and r, s are non-units then so is $r + s$;

(iv) the set of non-units of R is an ideal of R. //

In the commutative case, of course, a ring C is local if and only if it has a unique maximal ideal $\underline{m}$, and then the field $K = C/\underline{m}$ is called the residue field of C. If P is a prime ideal of a commutative ring C, and $S = C - P$ and M is a C-module, we shall write $S^{-1}M = M_P$ and $S^{-1}C = C_P$. In this case it follows easily from Proposition 2.3 that C_P is a local ring, with unique maximal ideal PC_P. This ring is called the localisation of C at P.

RING EXTENSIONS

Let A be a C-algebra and let the homomorphism $f : C \to A$ be injective. Then $LA_C(A) = RA_C(A) = 0$ since $1_A \in A$. We can then identify C with Cf and suppose that f is just set-theoretic inclusion. With this convention, we are in effect supposing that $C \subseteq Z(A) \subseteq A$. In this situation, an element $a \in A$ is said to be integral over C if there is an integer $n \geq 1$ and elements $c_1, \ldots, c_n$ of C such that

$$a^n + c_1 a^{n-1} + \dots + c_{n-1} a + c_n = 0 .$$

That is to say, a is integral over C if there is a monic polynomial $p(x) = x^n + c_1 x^{n-1} + c_n$ in $C[x]$, the polynomial ring in one central indeterminate over C, such that $p(a) = 0$.

A left (or right) modulet M over a ringlet R is said to be faithful if its annihilator is the zero ideal of R. Thus the situation at the start of this section was one in which the algebra A was a faithful module over the ring C.

2.5 Proposition. Let $R \subseteq S$ be commutative rings. The following conditions on $s \in S$ are equivalent:

(i) s is integral over R;

(ii) $R[s]$ is a finitely generated R-module;

(iii) $R[s]$ is contained in a subring C of S which is finitely generated as an R module;

(iv) there exists a non-zero faithful $R[s]$ module M which is finitely generated as an R-module.

Proof. (i) $\Rightarrow$ (ii). If $s^n + c_1 s^{n-1} + \dots + c_n = 0$ then $R[s] = R + Rs + Rs^2 + \dots + Rs^{n-1}$.

(ii) $\Rightarrow$ (iii). Let $R[s] = C$.

(iii) $\Rightarrow$ (iv). Let $C = M$.

(iv) $\Rightarrow$ (i). If $M = Rm_1 + \dots + Rm_t$ then for each i, $1 \leq i \leq t$,

$$sm_i = \sum_{j=1}^{t} a_{ij} m_j \qquad \text{for some } a_{ij} \in C,$$

$$\text{so} \quad \sum_{j=1}^{t} (a_{ij} - s\delta_{ij}) m_j = 0, \quad \text{for } 1 \leq i \leq t,$$

where $\delta_{ij} = 1$ or 0 according as $i = j$ or $i \neq j$. Since R is a commutative ring we may in the familiar manner treat this system as a single matrix equation, and then multiply on the left by the adjoint of the matrix $(a_{ij} - s\delta_{ij})$. This shows $\Delta m_j = 0$ for $1 \leq j \leq t$, where $\Delta = \det(a_{ij} - s\delta_{ij})$. Hence $\Delta M = 0$, so $\Delta = 0$ since M is faithful. But then $(-1)^t \Delta = 0$ is an equation of integral dependence of the type wanted. //

2.6 Corollary. Let $R \subseteq S$ be commutative rings and $a, b \in S$ be integral over R. Then so are $a + b$ and ab. Hence the set T of elements of S which are integral over R is a subring of S that contains R.

Proof. Since b is integral over R it is integral over $R[a]$, so $R[a, b]$ is a finitely generated $R[a]$ module; since $R[a]$ is a finitely generated R-module so is $R[a, b]$. Let $c = a + b$ and $d = ab$. Since $R[c] \subseteq R[a, b]$, and $R[d] \subseteq R[a, b]$, the first two results follow at once from part (iii) of the Proposition. The final sentence is then obvious. //

It is worth remarking that if $R \subseteq Z(S) \subseteq S$ for rings R and S, but S is not commutative, the results of Corollary 2.6 may fail. For example, let K be a commutative field and t a central indeterminate over K. Let R be the ring of 2×2 matrices $\{\begin{bmatrix} a & 0 \\ 0 & a \end{bmatrix} : a \in K\}$ and S be the full matrix algebra of 2×2 matrices over the polynomial ring $K[t]$. Let $x = \begin{bmatrix} 0 & t \\ 0 & 0 \end{bmatrix}$ and $y = \begin{bmatrix} 0 & 0 \\ t & 0 \end{bmatrix}$. Then $x^2 = y^2 = 0$ so each of x and y is integral over R, and R is certainly a central subring of S. But $xy = \begin{bmatrix} t^2 & 0 \\ 0 & 0 \end{bmatrix}$ and $yx = \begin{bmatrix} 0 & 0 \\ 0 & t^2 \end{bmatrix}$, and $x + y = \begin{bmatrix} 0 & t \\ t & 0 \end{bmatrix}$ so $(x + y)^2 = \begin{bmatrix} t^2 & 0 \\ 0 & t^2 \end{bmatrix}$. None of these elements is integral over R.

2.7 Lemma. Let $R \subseteq S$ be commutative rings and $u \in S$ be a unit of S. Then u is integral over R if and only if $u \in R[u^{-1}]$.

Proof. If u is integral over R there exist $a_1, \ldots, a_n \in R$ such that

$$u^n + a_1 u^{n-1} + \ldots + a_n = 0.$$

Multiplying by $u^{-(n-1)}$ we see that $u \in R[u^{-1}]$. Conversely, if $u \in R[u^{-1}]$, we have $u = -(a_1 + a_2 u^{-1} + \ldots a_n u^{-n+1})$ for some $a_1, \ldots, a_n \in R$, and then we immediately obtain an equation of integral dependence for u over R. //

If $R \subseteq S$ are commutative rings, the integral closure of R in S is the subring T of elements of S which are integral over R. If $S = T$ then S is said to be integral over R, and $R \subseteq S$ is called an integral extension. The following result is an immediate consequence (by induction) of Corollary 2.6.

2.8 Lemma. If $R \subseteq S$ are commutative rings and $s_1, \ldots, s_n$ are elements of S which are integral over $R[s_1, \ldots, s_n]$ is integral over R. //

We now prove an omnibus lemma containing various key results about an integral extension of commutative rings.

2.9 Lemma. Let $R \subseteq S$ be an integral extension of commutative rings, then:
(a) if I is an ideal of S then $J = I \cap R$ is an ideal of R and S/I is an integral extension of R/J;
(b) if R and S are integral domains then S is a field if and only if R is a field;
(c) if P is a prime ideal of S then $Q = P \cap R$ is a maximal ideal of R if and only if P is a maximal ideal of S.

Proof. (a) It is clear that J is an ideal of R, and that J is the kernel of the composite map $R \to S \to S/I$, so we have a ring extension $R/J \subseteq S/I$. If $s + I \in S/I$ there exist $a_1, \ldots, a_n \in R$ such that

$$s^n + a_1 s^{n-1} + \ldots + a_n = 0.$$

But then, if $b \to \bar{b}$ is the canonical map $S \to S/I$, we have

$$\bar{s}^n + \bar{a}_1 \bar{s}^{n-1} + \ldots + \bar{a}_n = 0$$

where $\bar{s} = s + I$ and $\bar{a}_i \in R/J$. Hence $R/J \subseteq S/I$ is an integral extension.

(b) Suppose $R \subseteq S$ are integral domains and S is a field. If $x \in R$, $x \neq 0$ then $x^{-1} \in S$ so $x^{-1} \in R[x]$ by Lemma 2.7; that is, $x^{-1} \in R$, so R is a field.

On the other hand, if $R \subseteq S$ are integral domains and R is a field, suppose $s \in S$, $s \neq 0$. Let n be the least positive integer such that

$$s^n + a_1 s^{n-1} + \ldots + a_n = 0$$

for some $a_1, \dots, a_n \in R$. If $a_n = 0$ then

$$s(s^{n-1} + a_1 s^{n-1} + \dots + a_{n-1}) = 0$$

so

$$s^{n-1} + a_1 s^{n-1} + \dots + a_{n-1} = 0$$

since S is an integral domain. But this contradicts the minimality of n, so

$$s(s^{n-1} + \dots + a_{n-1}) = -a_n \neq 0.$$

But $-a_n$ is a unit of the field R, so s is a unit of S; hence S is a field.

(c) In any extension $R \subseteq S$ of commutative rings, if P is a prime ideal of S then $Q = P \cap R$ is a prime ideal of R. The result then follows from (a) and (b). //

In the non-commutative case, if $R \subseteq S$ is an inclusion of rings and P is a prime ideal of S then $Q = R \cap P$ is an ideal of R, but need not be a prime ideal. For example, if S is the full algebra of 2×2 matrices over a commutative field K then $P = 0$ is the only prime ideal of S. If R is the subalgebra of upper triangular 2×2 matrices over K, the ideal A of R which consists of the strictly upper triangular matrices is nilpotent; in fact $A^2 = 0 \neq A$. Hence $R \cap P = 0$ is not a prime ideal of R.

Our main applications of these results are in the situation where $R = \mathbb{Z}$, the ring of integers, and S is the set of integral elements in a field F contained in the complex field $\mathbb{C}$. The ring of all elements of $\mathbb{C}$ integral over $\mathbb{Z}$ is called the <u>ring of algebraic integers</u>, and denoted by $\mathbb{O}$. If $x \in \mathbb{O}$ and x is a rational number then x is actually in $\mathbb{Z}$, as is easily verified.

An important relationship between integral extensions and rings of fractions is provided by the following result.

<u>2.10 Lemma.</u> Let $R \subseteq S$ be an integral extension of commutative rings, and let $M \subseteq R$ be a multiplicatively closed set. Then $M^{-1}R \subseteq M^{-1}S$ is an integral extension.

Proof. Observe first that the kernel of the composite map $R \to S \to M^{-1}S$ is simply the kernel of $R \to M^{-1}R$, namely $\{x \in R : \exists\, m \in M,\ xm = 0\}$. Hence $M^{-1}R \to M^{-1}S$ is injective, so we may think of it as a ring extension.

Let $s \in S$, $m \in M$. Since S is integral over R we can find elements $a_1, \ldots, a_n$ of R such that

$$s^n + a_1 s^{n-1} + \ldots + a_n = 0 \quad \text{in } S.$$

Hence $(s/m)^n + (a_1/m)(s/m)^{n-1} + \ldots + (a_n/m^n) = 0$ in $M^{-1}S$, and each $a_i/m \in M^{-1}R$. //

The next result is of fundamental importance.

2.11 Theorem. Let $R \subseteq S$ be an integral extension of commutative rings, and let P be a prime ideal of R. Then there exists a prime ideal Q of S such that $Q = R \cap P$.

Proof. Let $f : R \to R_P$ and $g : S \to S_P$ be the natural maps (in the notation explained after Proposition 2.4). By Lemma 2.10, S_P is integral over R_P. If B is a maximal ideal of S_P then by Lemma 2.9 $B \cap R_P$ is a maximal ideal of R_P, a local ring. Let $Q = g^{-1}(B) = \{x \in S : xg \in B\}$; then Q is a prime ideal of S by Proposition 2.3, and $Q \cap R = f^{-1}(B \cap R_P) = f^{-1}(PR_P) = P$, as required. //

CHAIN CONDITIONS AND TENSOR PRODUCTS

In this section we review briefly some classical results and techniques. We begin with the concept of composition series for a modulet. A modulet is simple if it is non-zero and has no non-trivial submodulets (equivalently, quotient modulets).

If M is a modulet over a ringlet R then a composition series for M is an ascending chain of submodulets

$$0 = M_0 \subset M_1 \subset M_2 \subset \ldots \subset M_n = M$$

such that, for $0 \leq i < n$, the factor modulet M_{i+1}/M_i is simple. The integer

n is the length of the composition series. If

$$0 = N_0 \subset N_1 \subset N_2 \subset \ldots \subset N_r = M$$

is another composition series for M, then the two series are said to be equivalent if $n = r$ and, for some permutation π of $\{0, \ldots, n-1\}$,

$$M_{i+1}/M_i \simeq N_{i\pi+1}/N_{i\pi}$$

There are two important results about composition series. We shall sketch the proof of each. The first is

2.12 Proposition. Let M be an R-modulet then M has a composition series if and only if M satisfies the ascending and descending chain conditions on submodulets.

Proof. Suppose the two conditions are satisfied, and $M \neq 0$. Then we can pick a minimal non-zero submodulet M_1, and $M_1/0$ is simple. If $M_1 \neq M$ then we can pick a submodulet M_2 minimal amongst the submodulets strictly containing M_1, and then M_2/M_1 is simple. Iterating, we will, by the a. c. c., find an integer n such that $M_n = M$, and then we have a composition series as required.

The reverse implication follows by induction on n, since we can assume that, if $n > 1$, M_{n-1} satisfies both chain conditions as does M/M_1. //

The other important result about composition series is

2.13 Proposition. Let M be an R-modulet then any two composition series for M are equivalent.

Proof. If the two series are

$$0 = M_0 \subset M_1 \subset \ldots \subset M_n = M \quad \text{and}$$

$$0 = N_0 \subset N_1 \subset \ldots \subset N_r = M$$

we proceed by induction on n. The case $n = 1$ is obvious, and then the case

$M_1 = N_1$ follows at once by induction. If $M_1 \neq N_1$ then $M_1 \cap N_1 = 0$, and there exists a least integer $s > 1$ such that $M_1 \subset N_s$. If $s = r$ then $M = M_1 \oplus N_{r-1}$ and if $s \leq r-1$ we can look separately at the modulets N_s and M/N_s. The remaining details are left to the reader. //

The above discussion involves results for modulets which closely resemble the results for groups which we mentioned after Corollary 1.18.

We now come to a generalisation of a classical result known as Schur's Lemma.

2.14 Proposition. Let R be a ringlet and M a (left) R-modulet.

(a) If M satisfies a.c.c. and has a unique maximal submodulet N then $E = \text{End}_R(M)$ is a local ring with maximal ideal $\underline{m} = \{f \in E : Mf \subseteq N\}$

(b) If M satisfies d.c.c. and has a unique minimal submodulet N then $E = \text{End}_R(M)$ is a local ring with maximal ideal $\underline{m} = \{f \in E : Nf = 0\}$.

(c) If M is simple then $D = \text{End}_R(M)$ is a division ring.

Proof. We prove (a); the proof of (b) is virtually identical (but with all inclusions reversed) and then (c) is a special case of either (a) or (b).

Let $f \in E$ and suppose $Mf = M$. We wish to show that f is injective, and hence a unit of E. Let K_i be the kernel $\text{Ker } f^i$ of f^i for $i = 1, 2, \ldots$. Clearly $K_1 \subseteq K_2 \subseteq \ldots \subseteq K_h \subseteq \ldots$. So there exists an integer n such that $K_n = K_{n+1} = \ldots = K_{2n}$, and it follows easily (comparing K_n and K_{2n}) that $K_n = 0$ so $K_1 = \text{Ker } f = 0$. On the other hand, $\underline{m} = \{f \in E : Mf \subseteq N\}$ is clearly a right ideal of E, so it is the unique maximal right ideal, and $E/\underline{m}$ is a division ring. //

The final statement of this Proposition is known as Schur's Lemma.

Let R be a ringlet and M, N be R-modulets, and let $\text{Hom}(M, N) = \text{Hom}_R(M, N)$ denote the set of R-homomorphisms from M to N. It follows from the definitions that if M, N are simple modulets either $\text{Hom}(M, N) = 0$ or $M \simeq N$, but not both. More generally, it follows easily from a consideration of composition series (Proposition 2.13) that if $M, N_1, \ldots, N_n$

are simple R-modulets then $\mathrm{Hom}(M, N_1 \oplus \ldots \oplus N_n) \neq 0$ if and only if $M \simeq N_i$ for some i, $1 \leq i \leq n$. Similarly, if $N_1, \ldots, N_n$ are isomorphic simple R-modulets and $D = \mathrm{End}(N_i)$ for each i, it follows that $\mathrm{End}(N_1 \oplus \ldots \oplus N_n) \simeq M_n(D)$, the full matrix algebra over D of $n \times n$ matrices.

The other classical technique which we shall need later is the tensor product. Although we are chiefly interested in the tensor products of modules over commutative rings, we begin with a somewhat more general construction, for one-sided modules over rings. We make no attempt, however, to consider the case of modulets over ringlets.

Suppose M, N are respectively a right and a left module over a ring R, and P is an additive abelian group. A mapping $f : M \times N \to P$ is said to be <u>balanced</u> if, for all $r \in R$ and $m_i \in M$, $n_i \in N$

$$f(m_1 + m_2, n_1) = f(m_1, n_1) + f(m_2, n_1)$$
$$f(m_1, n_1 + n_2) = f(m_1, n_1) + f(m_1, n_2)$$
$$f(m_1, rn_1) = f(m_1 r, n_1).$$

The first two conditions are sometimes described by saying that f is <u>$\mathbb{Z}$-bilinear.</u> The <u>tensor product</u> of M and N over R is written $M \otimes_R N$, and is the "unique" pair f, T whose existence and uniqueness are asserted by the following result.

<u>2.15 Theorem.</u> Let R be a ring and M, N be a right and a left R-module respectively. Then there exists an additive abelian group T and a balanced map $f : M \times N \to T$ such that

(i) T is generated by $\{f(m, n) : m \in M, n \in N\}$; that is, every element of T is a sum of elements of the form $f(m, n)$ for $m \in M$, $n \in N$;

(ii) If $p : M \times N \to P$ is a balanced map of $M \times N$ to an additive abelian group P then there exists a unique group homomorphism $\theta : T \to P$ such that $p = \theta f$;

(iii) If f', T' is another pair satisfying the two previous conditions then there is a unique group isomorphism $\phi : T \to T'$ such that $\phi f = f'$.

Proof. It is clear that (iii) follows from (ii). Now let F be the free abelian group generated by the pairs $(m, n) \in M \times N$ and let K be the subgroup of F generated by the elements

$$(m_1 + m_2, n_1) - (m_1, n_1) - (m_2, n_1)$$
$$(m_1, n_1 + n_2) - (m_1, n_1) - (m_1, n_2)$$
$$(m_1, rn_1) - (m_1 r, n_1).$$

Let T be the group F/K, let $f : M \times N \to T$ be the map defined by $f(m, n) = (m, n) + K$. It is easy to check that f is a balanced map, and then that, for any integer $g \in \mathbb{Z}$

$$f(mg, n) = f(m, gn) = gf(m, n)$$

so that T is generated in the required way. If $p : M \times N \to P$ is a balanced map then p has a unique extension to a group homomorphism say $p^* : F \to P$, and $K \subseteq \ker p^*$ since p is balanced. The existence and uniqueness of the group homomorphism $\theta : T \to P$ then follow at once. //

We shall denote f(m, n) by $m \otimes n$, or $m \otimes_R n$ where ambiguity is possible. If L is a submodule of M then, for $m \in L$, the notation $m \otimes n$ could refer either to $m \otimes n \in L \otimes N$ or $m \otimes n \in M \otimes N$; in practice it will be clear which is meant from the context. Note that there is, in general, no sensible way of including $L \otimes N$ in $M \otimes N$ so as to eliminate the ambiguity. For example, let $R = L = \mathbb{Z}$ be the ring of rational integers, let M be the R-module of rational numbers and let N be $\mathbb{Z}/n\mathbb{Z}$ for some integer $n > 1$. If $x \otimes y \in M \otimes N$ then $x \otimes y = x/n \,.\, n \otimes y = x/n \otimes ny = 0$. So $M \otimes N = 0$, but $L \otimes N \cong N \neq 0$.

Let M be an R - S bimodule and N be a left S module, then $M \otimes_S N$ can be given the structure of a left R-module. For the mapping $(m, n) \to rm \otimes n$ of $M \times N$ to $M \otimes_S N$ is balanced, for any $r \in R$, and hence gives rise, in the fashion of Theorem 2.15, to a group homomorphism $\theta = \theta_r : M \otimes_S N \to M \otimes_S N$ such that $\theta(\Sigma m_i \otimes n_i) = \Sigma rm_i \otimes n_i$ for all $m_i \in M$, $n_i \in N$. We define left multiplication by elements of R on $M \otimes_S N$ by

$$rx = \theta_r(x)$$

for all $r \in R$, $x \in M \otimes_S N$, and it is straightforward to check that with this definition $M \otimes_S N$ is a left R module. The right analogue of this result will also be used without further comment.

The most important case arises when S is commutative, and then each of M and N can be thought of as an S - S bimodule in which each element is centralising. Suppose that we use the procedures of the previous paragraph to give $M \otimes_S N$ the structure of a left S module, or alternatively the structure of a right S module. Then since S is commutative we can in each case extend the construction to give $M \otimes_S N$ the structure of an S - S bimodule in which every element is centralising. There is no difficulty in seeing that, whichever route we follow, we obtain in the end the same bimodule structure on $M \otimes_S N$.

Let S be a commutative ring and M, N be S modules (which we think of as bimodules in which every element is centralising). Then for a non-zero element $a \in S$ we have, for $(m, n) \in M \times N$,

$$a(m \otimes n) = (am) \otimes n = (ma) \otimes n = m \otimes (an) = m \otimes (na) = (m \otimes n)a$$

so, for $a_i \in S$, $m_i \in M$, $n_i \in N$,

$$(a_1 m_1 + a_2 m_2) \otimes n_1 = a_1(m_1 \otimes n_1) + a_2(m_2 \otimes n_1)$$
$$m_1 \otimes (a_1 n_1 + a_2 n_2) = a_1(m_1 \otimes n_1) + a_2(m_1 \otimes n_2).$$

These conditions are sometimes described by saying that the tensor product over a commutative ring S is S-bilinear.

THE JACOBSON RADICAL

In this section we give a brief development of the Jacobson radical of a ringlet, and use it to prove a famous theorem due to Artin and Wedderburn. Our treatment of the Jacobson radical is rather different from that commonly given, for example in Herstein's excellent book [14].

We begin with some definitions. A ringlet R is simple if $R \neq 0$ and the only ideals of R are itself and 0. A ringlet R is nilpotent if $R^n = 0$ for some $n \geq 1$. If R is simple and non-nilpotent then R is idempotent; that is, $R^2 = R$. If R is simple and nilpotent then $R^2 = 0$. It is clear that a simple nilpotent ringlet has p elements for some rational prime p.

If R is a ringlet and M is a simple (say) right R-modulet then either $RA(M) = R$ or $MR = M$ (but not both). It is clear that if $RA(M) = R$ and M is simple then M has p elements for some rational prime p. It is also clear that there are very close parallels between the properties of simple ringlets and of simple modulets.

A maximal right or left ideal M of a ringlet R is said to be regular if the one-sided modulet R/M is not annihilated by R, or equivalently if $R = R^2 + M$. Clearly M is regular if and only if $R^2 \not\subseteq M$, and M is not regular if and only if M is a two-sided ideal of R and R/M is a simple nilpotent ringlet.

The following Lemma gives an alternative characterisation of a regular maximal one-sided ideal.

2.16 Lemma. A maximal right ideal M of a ringlet R is regular if and only if there exists an element $e \in R$ such that $a - ea \in M$ for every $a \in R$.

Proof. Suppose M is regular. Then there exists an element x of the right R-modulet R/M such that $R/M = xR$, and $M = RA(x)$. Now $x \in xR = R/M$ so $x = xe$ for some $e \in R$. But then if $a \in R$ $xa = xea$ so $x(a - ea) = 0$ i.e. $a - ea \in M$.

Conversely, suppose there exists $e \in R$ with the given property. Clearly $e \notin M$. Let $x = e + M \in R/M$. Then $R/M = xR$, so $R = R^2 + M = eR + M$. //

The left analogue of this result will be used without further comment. Note that if M is an additive abelian group of order p, a rational prime, we can always define a right or a left R-modulet structure on M, for any ringlet R, by $MR = 0$ or $RM = 0$, respectively. Thus R is always the annihilator of a simple right and a simple left R modulet.

2.17 Lemma. Let R be a ringlet and $A \neq R$ be a right ideal. Suppose there is an element e of R such that $a - ea \in A$ for every $a \in R$. Then there is a regular maximal right ideal of R such that $R = eR + M$ and $A \subseteq M$.

Proof. The lemma will follow from 2.16 if we can show that there is a maximal right ideal $M \neq R$ containing A. Let Σ denote the set of right ideals B of R such that $B \supseteq A$ and $B \neq R$. Σ is not empty as $A \in \Sigma$. It is also easy to see that $e \notin B$ for otherwise $B = R$. By Zorn's Lemma there is a maximal element M of Σ. It is clear that M is a maximal right ideal of R. //

We now define the right Jacobson radical of a ringlet R to be the set of elements of R which annihilate every simple right R-modulet M, and denote it (temporarily) by $RJ(R)$. It is clear that $RJ(R)$ is the intersection of the annihilators of all the simple right R-modulets, and hence is a two-sided ideal. The left Jacobson radical $LJ(R)$ is analogously defined.

An element a of a ringlet R is said to be right quasi-regular if there exists $a' \in R$ such that $a + a' + aa' = 0$, and then a' is said to be a right quasi-inverse of R. The concepts left quasi-regular and left quasi-inverse are similarly defined. A one-sided ideal is right (or left) quasi-regular if each of its elements is.

2.18 Lemma. If R is a ringlet and $a \in R$ has right and left quasi-inverses a' and a'' respectively then $a' = a''$.

Proof. We have $a + a' + aa' = 0 = a + a'' + a''a$ so $a''a + a''a' + a''aa' = 0 = aa' + a''a' + a''aa'$, so $a''a = aa'$ and from the original equations it follows that $a' = a''$. //

Let $x \in R$ and let $B = \{a + xa;\ a \in R\}$. Observe that B is right ideal of R and that $x \in B$ if and only if there is a $y \in R$ such that $x = y + xy$. Then

$$x + (-y) + x(-y) = 0 .$$

So $x \in B$ if and only if x has a right-quasi-inverse. Also note that

$a - (-x)a \in B$ for all $a \in R$. So $B = R$ if and only if $x \in B$ which is true if and only if x has a right-quasi-inverse. If $B \neq R$ we can apply Lemma 2.17. These comments play a key role in the proof of (2.19).

2.19 Lemma. Let R be a ringlet. The $RJ(R)$ is a right quasi-regular ideal containing every right quasi-regular right ideal of R. Further, $RJ(R)$ is also a left quasi-regular ideal of R.

Proof. Let $x \in RJ(R)$, and let B be the right ideal $\{xa + a : a \in R\}$. If $x \in B$ then x is right quasi-regular. If $x \notin B$ then B satisfies the conditions of Lemma 2.17 (with $-x$ here replacing x there) so there is a maximal right ideal M of R such that $R = xR + M$ (since $xR = (-x)R$). But then $x \notin RA(R/M)$; contradiction. Hence $RJ(R)$ is right quasi-regular.

Suppose $x = xy + y \in B$; note that $-y = -x + (-x)y \in RJ(R)$ so $-y$ has a right quasi-inverse z. But x is a left quasi-inverse for $-y$, so by Lemma 2.18 $x = z$. Hence x has left quasi-inverse $-y$, so $RJ(R)$ is left quasi-regular.

Now suppose A is a right quasi-regular right ideal of R, and let $a \in A$. If $a \notin RJ(R)$ then there is a simple right R-modulet M such that $Ma \neq 0$. Clearly $M = wR$ for some $w \in M$ such that $wa \neq 0$. Now $N = \{x \in M : xR = 0\}$ is a submodulet of M, and $N \neq M$, so $N = 0$. Hence $waR \neq 0$, so $M = waR$. Thus $-w = was$ for some $s \in R$. But since $as \in A$ and A is right quasi-regular, there exists $t \in R$ such that $as + t + ast = 0$. It follows that $(-w)(as + t + ast) = -was - wt - wast = w - wt + wt = w = 0$; contradiction. Hence $a \in RJ(R)$. //

We now obtain the following result, characterising the Jacobson radical $J(R)$ (unqualified by "right" or "left") of a ringlet R. This characterisation is fundamental to modern ring theory.

2.20 Theorem. Let R be a ringlet. Then the right and left Jacobson radicals of R are equal. Further,

(i) $J = \cap A$, where A runs over the annihilators of simple right R-modulets;

(ii) $J = R \cap \cap M$, where M runs over the regular maximal right ideals (if any) of R;

(iii) J is the unique maximal right quasi-regular right ideal of R;

(iv) J is also characterised by the left analogues of each of (i), (ii) and (iii).

Proof. That RJ(R) = LJ(R) follows at once from Lemma 2.19 and its left analogue. Assertions (i) and (iii) then follow from the definitions and Lemma 2.19 again. The inclusion $J \subseteq R \cap \cap M$, where M runs over the regular maximal right ideals (if any) of R is immediate from the definition, as is the reverse inclusion in the case where R has no such right ideals. So suppose R has at least one regular maximal right ideal, and let I denote the intersection of all of them. We need to show $I \subseteq J$. So suppose $x \in I$, $x \notin J$. Then $x \notin \{xy+y : x \in R\}$, and as in the proof of Lemma 2.19 there exists a regular maximal right ideal N such that $R = xR + N$; contradiction. The left analogues of (i), (ii), (iii) are then clear. //

It is worth dwelling further on the parallels between the Jacobson radical of a ringlet and the Frattini subgroup of a group. These parallels are highlighted by the formal similarities between Theorem 2.20 and Proposition 1.23, but are most clearly seen in the case of a ring with unity.

We first, however, note that in some cases it is easy to tell that an element or a one-sided ideal of a ringlet is in its Jacobson radical. An element x of a ringlet is <u>nilpotent</u> if $x^n = 0$ for some integer $n \geq 1$, and a one- or two-sided ideal is <u>nil</u> if every element of it is nilpotent.

2.21 Proposition. Let A be a nil one-sided ideal of a ringlet R. Then A is contained in the Jacobson radical J = J(R) of R.

Proof. If $x \in A$ and $x^n = 0$, let $y = \sum_{1}^{n-1} (-x)^j$ then $x + y + xy = 0 = x + y + yx$. Hence A is a right quasi-regular right ideal or a left quasi-regular left ideal, and in any case is contained in J(R) by Theorem 2.20. //

Suppose R is a ring with unity. Then every maximal one-sided ideal is regular, since $R = R^2$, and (by Zorn's Lemma) every right (left) ideal $N \neq R$ is contained in a maximal right (left) ideal. If S is a ringlet and M is an S modulet, a submodulet T of M is superfluous if, whenever N is a submodulet of M, if $M = N + T$ then $M = N$. An element x of M is superfluous if the submodulet which it generates is superfluous. The concept of a superfluous element of a modulet is analogous to that of a non-generator of a group. If R is a ringlet then an element x of R is right superfluous (or left superfluous) when the right ideal (or the left ideal) which it generates is a superfluous submodulet of the right R-modulet R_R (or the left R-modulet $_RR$). In the case of a ring, there is no need to distinguish between a right superfluous and a left superfluous element, as the following result shows.

2.22 Theorem. Let R be a ring with Jacobson radical J. The following conditions on an element x of R are equivalent:

(1) $x \in J$

(2) x is right superfluous

(3) x is left superfluous.

Proof. Suppose $x \in J$ and $R = N + xR$ for some right ideal N of R. If $R \neq N$ then there is a maximal right ideal M of R which contains N, and then $R = M + xR$. But $J \subseteq M$, so $R = J$, which is impossible.

Conversely, suppose x is right superfluous, and $x \notin J$. Then there exists a maximal right ideal M such that $x \notin M$. But then $R = M + xR$ so x is not right superfluous; contradiction. The equivalence of (1) and (3) follows similarly. //

It should be pointed out that, with the definitions here, Theorem 2.22 can fail for a ringlet without unity. Let A be an additive abelian group of order p (so that A is cyclic, generated by x, say). Define a ringlet structure on A by $A^2 = 0$. By Proposition 2.21, $J(A) = A$ and clearly $J(A)$ is the right ideal generated by x. But $A = 0 + J$ and $A \neq 0$, so Theorem 2.22 does not extend to the case of A.

TWO CLASSIC THEOREMS

Let R be a ringlet. We say that R is a semiprime ringlet if $R \neq 0$ and, for each ideal I of R, $I^2 = 0$ implies $I = 0$. A trivial inductive argument shows that, if R is a semiprime ringlet and I is an ideal such that $I^m = 0$ for some integer $m \geq 1$, then $I = 0$. An ideal N of R is said to be semiprime if R/N is a semiprime ringlet. Clearly a prime ideal is semiprime. We can now prove together two classic results due to Wedderburn and Artin.

2.23 Theorem. Let R be a semiprime ringlet which satisfies the d.c.c. on right ideals. Then R satisfies both the a.c.c. and the d.c.c. on right and on left ideals. Furthermore, there exist positive integers $m, n_1, \ldots, n_m$ and division rings D_i such that $R = \bigoplus_1^m M_{n_i}(D_i)$ is a direct sum of ideals each of which is the full matrix algebra $M_{n_i}(D_i)$ of $n_i \times n_i$ matrices over D_i. The integer m is unique and the integers $n_1, \ldots, n_m$ and the division rings $D_1, \ldots, D_m$ are unique up to permutation.

Conversely any such ring is a semiprime ringlet with d.c.c. on right ideals. Finally R is a prime ringlet if and only if $m = 1$, or equivalently if and only if R is a simple ring.

Proof. (a) We first show that R can be written as a direct sum of ideals each of which is a direct sum of isomorphic right ideals.

Choose ${}^1I$ to be a minimal right ideal of R. Let Σ denote the set of direct sums of minimal right ideals of R each of which is isomorphic to ${}^1I$. Since R has d.c.c. on right ideals Σ contains no infinite direct sums so we can find minimal right ideals $I_1, \ldots, I_n$ such that ${}^1I \simeq I_j$ for $1 \leq j \leq n$, the sum $I_1 + \ldots + I_n$ is direct, and if I is a minimal right ideal and ${}^1I \simeq I$ then $I \subseteq I_1 \oplus \ldots \oplus I_n$. If H is a minimal right ideal and $y \in R$ then yH is either zero or a minimal right ideal, isomorphic to H (with the isomorphism $h \to yh$). Hence $I_1 \oplus \ldots \oplus I_n$ is a non-zero two-sided ideal, say $P({}^1I)$. If $R = P({}^1I)$ we can stop here; if not then R has a minimal right ideal ${}^2I$ such that ${}^2I$ is not isomorphic to ${}^1I$, and we may similarly construct an ideal $P({}^2I)$ which is a finite direct sum of minimal right ideals each isomorphic to ${}^2I$. By the remarks following Proposition 2.14, $P({}^1I) \cap P({}^2I) = 0$. If $R \neq P({}^1I) \oplus P({}^2I)$, we may repeat the construction, to obtain a sequence ${}^1I, {}^2I, \ldots$ of pairwise non-isomorphic minimal right ideals such that the sum $P({}^1I) \oplus P({}^2I) \oplus \ldots P({}^mI) \oplus \ldots$ is direct. Since R certainly has no infinite direct sums of two-sided ideals, we must have $R = P({}^1I) \oplus \ldots \oplus P({}^mI)$ for some integer $m \geq 1$.

Py Proposition 2.14 and the remarks following it, it follows that $End_R(P({}^jI)_R)$, the ring of right R-modulet endomorphisms of $P({}^jI)$, is isomorphic to the full matrix algebra $M_{n_j}(D_j)$ of $n_j \times n_j$ matrices over a division ring D_j, where $D_j \simeq End({}_jI)$ and n_j is the length of a right composition series of $P({}^jI)$; R, and each $P({}^jI)$, clearly have right composition series.

(b) We show that if $x \in R$ and $x \neq 0$ then $xR^i \neq 0$ and $Rx^i \neq 0$. Assume that $xR = 0$. Let $\langle x \rangle$ denote the additive group generated by x then if I is the ideal generated by x, $I = \langle x \rangle + Rx + xR + RxR$. As $xR = 0$ $I = \langle x \rangle + Rx$. Then $I^2 = (\langle x \rangle + Rx)(\langle x \rangle + Rx) = 0$. Thus $I = 0$ and since $x \in I$, $x = 0$. We can argue similarly if $Rx = 0$ or $(xR)^2 = 0$.

(c) We now show that, for each i, there exists $u = u_i \in P({}^iI)$ such that $u = u^2$ and $P({}^iI) = P({}^iI)u = uP({}^iI)$. We need consider only the case of $P = P({}^1I) = I_1 \oplus \ldots \oplus I_n$, where $n = n_1$ and each $I_j \simeq {}^1I$, $1 \leq j \leq n$. If xR is a minimal right ideal of R there exists $y \in xR$ such that $yR = y(yR) = xR \neq 0$. But since $y \in yR = y^2R$, there exists $e \in yR$ such that $y = ye$, so $y(e^2 - e) = 0$, so $e^2 - e \in RA(y) \cap yR = 0$. Hence $e = e^2$, and for each i we can find $e_i \in I_i$ such that $0 \neq e_i = e_i^2$, so $e_iR = I_i$; let $u = \sum_1^n e_i$.

Now suppose $e \in P({}^1I)$ and $e = e^2$ and eR is a right ideal of R. If $x \in R$ then $x = x - ex + ex$ and $e(x - ex) = 0$ so $R = RA(e) + eR$. We know that $RA(e) \cap eR = 0$, so the sum is direct. Similarly $P({}^1I) = \{RA(e) \cap P({}^1I)\} \oplus eR$. It follows that $i \neq j$ implies $e_ie_j = 0$, and hence $u = u^2$ and $uP({}^1I) = P({}^1I)$. Now R is semiprime, so if A, B are ideals of R then $AB = 0$ if and only if $A \cap B = 0$ (because if $C = A \cap B$ and

$AB = 0$ then $C^2 = 0$). Hence $LA(P(^1I)) = RA(P(^1I)) = LA(uP(^1I)) = LA(uR) = LA(u) = \bigoplus_{j=2}^{n} P(^jI)$. If $z \in R$ then $z = z - zu + zu$, so $z \in Ru + \bigoplus_{j=2}^{n} P(^jI)$, and since $Ru \subseteq uR = P(^1I)$ we have $Ru = uR = P(^1I)u = uP(^1I)$, as required.

(d) We can now complete the proof of the first part of the theorem.

It has been shown that for each i there exists a u_i such that $u_i^2 = u_i$ and $P(^iI) = P(^iI)u_i = u_iP(^iI)$. Since $R = \bigoplus P(^iI)$ we also have that $P(^iI)P(^jI) = 0$ if $i \neq j$, and $u_iu_j = 0$ if $i \neq j$. Let $x \in P(^iI)$ and let $y \in P(^iI)$ such that $x = uy$. Then $ux = u^2y = uy = x$. Let α be any right R-endomorphism of $P(I)$. Then $\alpha(x) = \alpha(ux) = \alpha(u)x$. So any such endomorphism can be represented by left multiplication. Hence $End(P(^iI)_R)$ is isomorphic to $P(^iI)_R$. Since we showed earlier that $P(^iI)_R \cong M_{n_i}(D_i)$ for $D_i = End_R(I)$ the results follow using dimension arguments and the direct decomposition.

It remains to show that R is a simple ring if and only if R is a prime ringlet. This follows from the fact that, if R has more than one isomorphism class of minimal right ideal then R has at least two non-zero ideals A, B such that $A \cap B = 0$, while R has only one such isomorphism class if and only if R is isomorphic to the full matrix algebra $M_n(D)$ over a division ring D. //

SEMISIMPLE RINGS AND MODULES

The rings described in Theorem 2.23 are called semi-simple rings. If R is a ring and M is a one-sided R-module then M is said to be semi-simple if it satisfies any of the equivalent conditions of the following proposition.

2.24 Proposition. The following conditions on a right (or left) module M over a ring R are equivalent:

(i) M is the sum of its simple submodules;

(ii) M is a direct sum of a family of simple submodules;

(iii) every submodule of M is a direct summand of M.

Proof. If (i) is satisfied then by Zorn's Lemma M contains a maximal direct sum of simple submodules, say L. If $L \neq M$ then there is a simple submodule

N such that $N \subsetneq L$, whence $L \cap N = 0$, so the sum $L + N$ is direct; contradiction. Hence (ii) is satisfied.

If (ii) is satisfied and K is a submodule of M, suppose $K \neq M$. If $M = \bigoplus_{i \in I} N_i$ where each N_i is simple, we can find a subset J of I (using Zorn's Lemma) such that J is maximal with respect to $K \cap \bigoplus_{j \in J} N_j = 0$. As in the previous paragraph, we deduce that $K + \bigoplus_{j \in J} N_j = M$ since it contains every N_i, $i \in I$. Hence (iii) is satisfied.

Now suppose (iii) is satisfied, and let H be a submodule of M. We now show that, if $H \neq 0$ then H contains a simple submodule of M. Clearly we may suppose H is cyclic, and for the sake of definiteness we suppose M is a right R-module. If $H = mR$ where $m \in H$, then $RA(m) = E \neq R$ is a right ideal of R, contained in a maximal right ideal F. There exists a submodule G of M such that $M = mF \oplus G$. If $m = mf + g$ where $f \in F$ and $g \in G$ then $g = m(1 - f)$. Hence $gR = m(1 - f)R \subseteq mR \cap G$. But mR/mF is simple, so gR is a simple submodule of H. If D is the sum of all the simple submodules of M, then $M = D \oplus D'$ for some submodule D' of M. If $D' \neq 0$ then D' must contain a simple submodule of M, which contradicts the definition of D. Hence $M = D$ and (i) is satisfied. //

The connection between semisimple rings and semisimple modules is explained by the following theorem.

<u>2.25 Theorem.</u> Let R be a ring. Then the following conditions on R are equivalent

(1) R is semi-simple

(2) every right R-module is semi-simple

(3) every left R-module is semi-simple.

<u>Proof.</u> We will show that (1) and (2) are equivalent: the proof that (1) and (3) are equivalent is virtually identical.

Suppose first that R is semi-simple. It follows from the proof of Theorem 2.23 that R is a finite direct sum of simple right R-modules. It suffices by

Proposition 2.24 to show that a cyclic right R module xR is a direct sum of simple right R modules. If L is the right ideal $RA(x)$ of R then as in the proof of (2.24) there is a (finite) direct sum L' of simple right submodules of R such that $R = L \oplus L'$. But then $xR = xL' \simeq L'$, so xR is a direct sum of simple right R-modules.

Conversely, suppose every right R-module is a direct sum of simple R-submodules; then certainly the right R-module R_R is such a direct sum. There exists a finite collection of simple direct summands, say $I_1, \ldots, I_n$, such that $1 \in I_1 \oplus \ldots \oplus I_n$. But then $x \in I_1 \oplus \ldots \oplus I_n$ for any $x \in R$, so $R = I_1 \oplus \ldots \oplus I_n$. Thus R has a right composition series, and the Jacobson radical $J(R)$ of R is clearly zero, since it excludes each I_j. Hence it suffices to show that a ring in which $J(R) = 0$ is semiprime. Let R be such a ring, and A a non-zero ideal of R. There exists a maximal right ideal N such that A does not annihilate the simple right module $K = (R/N)$. Since $KA \neq 0$ we have $K = KA = KA^2$, so $A^2 \neq 0$. //

TWO LEMMAS ON INTEGRAL DOMAINS

Recall that an integral domain is a non-zero ringlet in which the product of two non-zero elements is non-zero. An element e of a ringlet R is idempotent if $e = e^2$; we met such elements in the proof of Theorem 2.23. In this section we prove two elementary lemmas concerning idempotents in integral domains.

2.26 Lemma. If D is an integral domain and $e \neq 0$ is an idempotent of D then e is a unity element for D.

Proof. If $x \in D$ then $(x - xe)e = 0$ and $e(x - ex) = 0$ so $x - xe = x - ex = 0$; that is, e is a unity element for D. //

2.27 Lemma. If D is an integral domain and C is a non-zero subringlet of D which satisfies right d.c.c. then C is a division ring, D has a unity element, and the unity elements of C and D coincide.

Proof. Let $0 \neq c \in C$. There exists an integer $r > 0$ such that $c^r C = c^{r+1} C$, so $c^{r+1} = c^{r+1} e$ for some $e \in C$. But then $c^{r+1}(e - e^2) = 0$, so $e \neq 0$ is an idempotent of D contained in C, and it remains, using Lemma 2.26, only to show that C is a division ring. But since $e \in C$ we have $c^r = c^{r+1} d$ for some d, and then $c^r(e - cd) = 0$ so $e = cd$. But $c(dc - e) = ec - ce = 0$, since e is a unity element, so $dc = e$, as required. //

REMARK

There are in the literature many not necessarily compatible uses of the term semisimple. One class of ambiguities comes from the theory of radicals (see [6]). In this approach, a radical F is a function on the class of ringlets that satisfies certain axioms, and then a ringlet R is F-semisimple if $F(R) = 0$. The Jacobson radical satisfies the axioms, and if $J(R) = 0$ then R is said to be J-semisimple. A more modern and (possibly) more common term for the same property is semiprimitive. A semiprimitive ringlet need not be semisimple in the sense used here, but it must be semiprime. It therefore follows that a ringlet is semisimple if and only if it is semiprimitive and has d. c. c. on right (equivalently, left). This is often taken as the definition of semisimple for the purposes of representation theory, because (classically) that theory is concerned with algebras which are finite-dimensional over a field.

3 Commutative algebra

INTRODUCTION

In this chapter we concentrate on commutative algebra. The material is divided into three sections all of which play an important role in later chapters. We begin with an introduction to Galois Theory where we emphasise the role the group plays. This view fits exactly with the approach taken in Chapter 1. This is a good place to recall that throughout this book field always means a commutative field.

The second section is devoted to an introduction of rings of algebraic numbers. We prove the results needed for the applications to representation theory. The approach is functional but we hope useful. The final section gives a proof of a result concerning primitive divisors, a concept which will be explained later. This theorem is variously attributed to Birkhoff, Vandiver and Zsigmondy. The proof we give is due to H. Luneburg [19].

GALOIS THEORY

Let F be a field and let $F[x]$ be the ring of polynomials over F. Recall that $F[x]$ is a commutative principal ideal domain with well-behaved factorization. Let $f(x)$ be an irreducible polynomial. Then the ideal generated by $f(x)$, $(f(x))$, is a maximal ideal of $F[x]$, and so $F[x]/(f(x))$ is a field, say E. Now F is naturally embedded in $F[x]$, and we can map $F[x]$ onto E by the natural map $g(x) \to g(x) + (f(x))$. It is easy to check that the combination of these two maps embeds F into E.

In general if $g(x) \in F[x]$ and $F \subseteq E$ is some ring extension we say that $\alpha \in E$ is a root of g if $g(\alpha) = 0$. When E is a (commutative) field then because $E[x]$ is a principal ideal domain it follows, using the well-behaved factorisation, that a polynomial $f(x) \in F[x]$ of degree n has at most n roots in E. In the case of the above construction notice that $x + (f(x))$ is a root of $f(x)$ in E. Thus given any irreducible polynomial $f(x)$ over a field F there exists an extension field E such that E contains a root of f.

If $F \subseteq E$ is a field extension and $u \in E$ we denote by $F(u)$ the smallest subfield of E containing u and F. There is a ring homomorphism Δ: $F[x] \to E$ such that $x\Delta = u$ and $f\Delta = f$ for every f in F. Thus $(a_0 + a_1 x + \dots + a_n x^n)\Delta = a_0 + a_1 u + \dots + a_n u^n$ for any integer $n \geq 1$ and any $a_0, \dots, a_n \in F$. If $f(x)\Delta = 0$ and $(f(x))$ is a maximal ideal of $F[x]$, it follows that $F[x]/(f(x)) \simeq F(u)$.

If $F \subseteq E$ is a field extension then $[E : F]$ will denote the dimension of E over F on an F-vector space. This is called the degree of the extension. The degree is multiplicative (as is worth checking) and we see from above that $[F[x]/(f(x)) : F] =$ degree of $f(x)$. If $[E : F]$ is finite then E is called a finite extension of F. Most of the comments we make here will concern finite extensions.

3.1 Lemma. Let $\alpha : F \to H$ be an isomorphism of fields. Let $f(x)$ be an irreducible polynomial in $F[x]$ and let $p(t)$ be the corresponding polynomial in $H[t]$, noting that α has an extension to a ring isomorphism $\alpha^* : F[x] \to H[t]$ such that $x\alpha^* = t$.

Let u be a root of f in a field $G \supseteq F$ and v be a root of p in a field $K \supseteq H$. Then there exists a unique isomorphism $\underset{\sim}{\beta}$ from $F(u)$ to $H(v)$ such that

(i) $\underset{\sim}{\beta}|_F = \alpha$ and

(ii) $u\underset{\sim}{\beta} = v$.

Proof. The homomorphism α^* is defined by putting $x\alpha^* = t$, $a\alpha^* = a\alpha$ for $a \in F$. To check that α^* is a well-defined ring isomorphism is not difficult. Under this isomorphism $(f(x)\alpha^* = p(t)$ and then we can define $\beta: F[x]/(f(x)) \to H(t)/(p(t))$ by using the natural maps. //

We shall in future, by abuse of notation, denote the extension α^* that was discussed in the above Lemma simply by α. The image of $(\sum_0^n a_i x^i)$ under α is simply $\sum_0^n (a_i\alpha) t^i$, so this is a very natural convention.

These ideas can be extended to arbitrary polynomials. Let $f(x) \neq 0$ be any polynomial in $F[x]$. If $f(x)$ is factorised into the product of irreducibles then by choosing any of them we can construct a field E_1 such that f has a root, say u_1, over E_1. Then consider $f(x)/(x_1 - u_1) = f_1(x)$, say, and construct E_2 so that f_1 has a root over E_2. This process can be repeated until there is a field, say E, such that $f(x)$ is a product of linear factors over E. If E is such a field, and if there is no smaller field E_0 such that $F \subset E_0 \subset E$ and f splits into linear factors over E_0, then E is called a splitting field for f over F.

We now come to an important theorem about splitting fields.

3.2 Theorem. Let F and H be two fields and let α be an isomorphism between them. Let f be a polynomial over F and let h be the corresponding polynomial over H. Let E and K be splitting fields for f over F and h over H respectively. Then there is an isomorphism $\gamma : E \to K$ such that $\gamma|_F = \alpha$ ("γ extends α").

Proof. We prove the result by induction on the degree of E over F. If $[E : F] = 1$ then there is nothing to prove since f splits into linear factors over F and hence splits into linear factors over H.

Let f_0 be an irreducible factor of f of degree >1, and let h_0 be the corresponding factor of h. Let u be a root of f_0 and v be a root of h_0. From (3.1) there exists an isomorphism β from $F(u)$ to $H(v)$ which extends α. But now E is a splitting field for f over $F(u)$ and $[E : F(u)] < [E : F]$. Also K is a splitting field for h over $H(v)$. So by induction there is an isomorphism γ from E to K which extends B, and hence α, as required. //

3.3 Corollary. If f is a polynomial over a field F then any two splitting fields of f are isomorphic. //

We now wish to introduce the concept of the Galois group of an extension E/F where $F \subseteq E$ are both fields. We shall develop only the Galois theory we need for the later results. Let E be a field and let $\mathrm{Aut}(E)$ be the group of all field automorphisms of E (or equivalently ring automorphisms). Let F be a subfield of E. Then

$$\mathrm{Gal}(E/F) = \{\alpha \in \mathrm{Aut}\, E \mid x\alpha = x \ \forall x \in F\}.$$

Since Gal(E/F) is defined to be that subset of a group the elements of which fix the points in a non-empty subset, clearly Gal(E/F) is a subgroup of Aut(E). We shall denote Gal(E/F) by G_F. Note that if $P \subseteq F \subseteq E$ are fields then G_F is a subgroup of G_P.

Before proceeding to the next theorem we need to make some comments about roots of polynomials and elements in extension fields. If $F \subseteq E$ is an extension of fields, an element u of E is said to be <u>algebraic over</u> F if it is the root of a non-zero polynomial in $F[x]$. For example, if $v \in F$ then v is the root of $x - v$. Suppose $w \in E$ and w is the root of the non-zero polynomial $p(x) = p_0 + p_1x + \ldots + p_nx^n$ where $p_0, \ldots, p_n \in F$ and $p_n \neq 0$. Then w is also the root of $p^*(x) = p_n^{-1}p(x)$, since $p_n(p^*(w)) = p_0 + p_1w + \ldots + p_nw^n = 0$, so $p^*(w) = 0$ since p_n is a non-zero element of the field E. But $p^*(x)$ has leading coefficient 1. Hence $u \in E$ is algebraic over F if and only if u satisfies a monic polynomial over F, or in other words if and only if u is integral over F.

Now suppose that u is an element of E algebraic over F, and that $f(x) \in F[x]$ is a monic polynomial of minimal degree amongst those of which u is a root. If $g(x)$ is another monic polynomial of the same degree and u is a root of $g(x)$ then u is also a root of $p(x) = f(x) - g(x)$. If $p(x) \neq 0$, then, as in the previous paragraph, we can find a monic polynomial $p^*(x) = ap(x)$, where $a \in F$, such that u is a root of $p^*(x)$. But clearly, in this event, $\deg p^*(x) < \deg f(x)$, which contradicts the choice of $f(x)$. Hence $f(x)$ is the only monic polynomial of degree $\deg f(x)$ which has u as a root. We shall say that the monic polynomial of minimal degree amongst those satisfied by u is the <u>minimum polynomial of u over F.</u>

Suppose $f(x) = g(x)h(x)$ in $F[x]$. Then $g(u)h(u) = 0$ in E, so $g(u) = 0$ or $h(u) = 0$. From the two previous paragraphs, we must therefore have $\deg g(x) = \deg f(x)$ or $\deg h(x) = \deg f(x)$, so that one of $g(x)$, $h(x)$ belongs to F. Thus $f(x)$ is an irreducible polynomial in $F[x]$, so $(f(x))$ is a maximal ideal of $F[x]$. Since $F(u) \simeq F[x]/(f(x))$, $\deg f(x) = [F(u) : F]$.

If $F \subseteq E$ is a finite extension of fields and $u \in E$ then u is algebraic over F, because the set $\{1, u, u^2, \ldots\}$ cannot be linearly independent.

3.4 Theorem. Let $F \subseteq E$ be an extension of fields. Let $G = \mathrm{Gal}(E/F)$ and let M, L be fields such that $F \subseteq L \subseteq M \subseteq E$ and that $[M : L] = n < \infty$. Then $[G_L : G_M] \leq n$.

Proof. We prove this result by induction on n. If $n = 1$ the result is obvious. So we assume the result holds for any $m < n$. Let K be a field such that $L \subset K \subset M$, if one exists. So $[M : K]$ and $[K : L]$ are both less than n. Using the inductive hypothesis,

$$[G_K : G_M] \leq [M : K] \qquad \text{and}$$
$$[G_L : G_K] \leq [K : L].$$

Since indices are multiplicative we obtain

$$[G_L : G_M] \leq [M : L] .$$

Thus we may assume that there is no proper subfield lying between M and L. Let $u \in M$ with $u \notin L$. Then $L(u) = M$ and so if f is a minimal polynomial for u over L, degree of f over L is n. Let $[G_L : G_M] = t$ and let $\alpha_1, \ldots \alpha_t$ be a set of right coset representatives for G_M in G_L. Now consider the set of images of u under the α_i s.

Because each $\alpha_i \in G_L$, we have $f . \alpha_i = f$ for each i. Thus each $u\alpha_i$ is a root of f. Assume that $u\alpha_i = u\alpha_j$ for some pair α_i, α_j. Then $u(\alpha_i\alpha_j^{-1}) = u$ and so $\alpha_i\alpha_j^{-1}$ fixes u. However $\alpha_i\alpha_j^{-1} \in G_L$ and so $\alpha_i\alpha_j^{-1}$ fixes u and L. Thus $\alpha_i\alpha_j^{-1}$ fixes $L(u)$ and so is in G_M. Hence $\alpha_i\alpha_j^{-1} = 1$.

We have now shown that $u\alpha_1, \ldots, u\alpha_t$ are all distinct and are all roots of f. Since f has degree n we conclude that $t \leq n$ as required. //

3.5 Corollary. If $F \subseteq E$ and $[E : F] = n$ then $|\mathrm{Gal}(E/F)| \leq n$.

Proof. Apply Theorem 3.4 with $M = E$ and $L = F$. //

We now introduce a new definition. If $|\mathrm{Gal}(E/F)| = n$ when $[E:F] = n < \infty$ then E is called a Galois extension of F. The significance of Galois extensions is explained by an important theorem involving their properties as splitting fields.

3.6 Theorem. Let E be a Galois extension of F. Let $f(x) \in F[x]$ be irreducible of degree m, and let $u \in E$ be a root of $f(x)$ over E. Then $f(x)$ has m distinct roots in E.

Proof. Let $[E : F] = n$ and let $G = \mathrm{Gal}(E/F)$. Consider $U = \{ug | g \in G\}$. Each element of U is a root of $f(x)$. The number of elements in U is $[G : H]$ where H is the subgroup of G which fixes u. But then H is the pointwise stabilizer of the field $F(u)$, so $H = G_{F(u)}$. Now $[F(u) : F] = m$ since $f(x)$ is irreducible of degree m, and so by (3.4), $[G : G_{F(u)}] \leq m$ and $[G_{F(u)} : 1] \leq n/m$. But since $|G| = n$ we must have $[G : G_{F(u)}] = m$ and $|G_{F(u)}| = n/m$. Thus U has m distinct elements and the result follows. //

3.7 Corollary. If E is a Galois extension of F then E is the splitting field of a polynomial f which has $\deg f$ distinct roots.

Proof. Let $[E : F] = n$ and let $u_1, \ldots, u_n$ be a basis of E over F. Let $f_1, \ldots, f_n$ be the minimum polynomials of $u_1, \ldots, u_n$ respectively. Let $g_1, \ldots, g_r$ be the distinct polynomials amongst the f_i. By (3.6) each of the g_i has distinct roots and each is irreducible. Thus no two can have a common root and so the polynomial $f = g_1 \ldots g_r$ is the one required by the corollary. //

We now come to the converse of this result, one of the cornerstones of Galois theory.

3.8 Theorem. Let F be a field and let f be a polynomial over F. Let E be a splitting field for f and assume that f has $\deg f$ distinct roots in E. Then E is a Galois extension of F.

Proof. We prove this result by induction on $[E : F]$. Note that if $[E : F] = 1$ there is nothing to prove. So suppose $[E : F] > 1$. Since $E \neq F$ there is a $u \in E$, $u \notin F$ such that u is a root of f. Let $F_1 = F(u)$. Note that E is still a splitting field for f over F_1. Then $[E : F_1] < [E : F]$ and since f will still have deg f distinct roots in E when E is regarded as an extension of F_1, by induction $\text{Gal}(E/F_1) = [E : F_1]$ because E is a Galois extension of F_1.

Let g be the minimum polynomial of u over F and let g have degree m. Then $F_1 = F(u)$ has degree m over F. Since u is a root of both f and g and g is irreducible g divides f. Thus g has deg g distinct roots, let us say $u = u_1, \ldots, u_m$. By the argument (3.2) we can extend the isomorphism from $F(u) \to F(u_j)$ given by (3.1) to an isomorphism $\alpha_j : E \to E$ which fixes F and maps u to u_j.

Thus each $\alpha_j \in G = \text{Gal}(E/F)$ and each α_j is in a distinct coset of $G_{F_1} = \text{Gal}(E/F_1)$. Thus $|G| = m|G_{F_1}| = m[E : F_1] = [E : F]$. So E is a Galois extension of F. //

<u>3.9 Corollary.</u> Let $F \subseteq L \subseteq E$ be an extension of fields. If E is a Galois extension of F then E is a Galois extension of L. //

Let $F \subseteq E$ be a field extension, and $f(x) \in F[x]$ be an irreducible polynomial which can be factorised into linear factors in $E[x]$. Then f(x) is said to be a <u>separable</u> irreducible polynomial if it has deg f distinct roots in E. (We shall shortly see that this definition is in fact independent of the field E.) A polynomial $p(x) \in F[x]$ is called <u>separable</u> if each of its irreducible factors is separable. If $u \in E$ and u is algebraic over F then u is said to be a <u>separable</u> element over F if its minimum polynomial over F is separable. Finally, if every element in E is algebraic and separable over F then E is said to be a <u>separable</u> (sometimes <u>separable</u> <u>algebraic</u>) extension of F.

Given a polynomial $f(x) = a_0 + a_1x + \ldots + a_nx^n$ we can define a formal derivative $f \to f'(x) = a_1 + 2a_2x + \ldots + na_nx^n$. This satisfies the usual rules for differentiation, that is, $(a+b)' = a' + b'$ and $(ab)' = a'b + ab'$. As a consequence if $f(x) = (g(x))^2h(x)$ then $f'(x) = 2g(x)g'(x)h(x) + g^2(x)h'(x)$. Thus g(x) divides both f(x) and f'(x). Using this observation it can be shown f(x) has no repeated roots if and only if f(x) and f'(x) have highest common factor 1.

Suppose f(x) is irreducible and let f(x), f'(x) have a common factor other than 1. Since $\deg f'(x) < \deg f(x)$ and f(x) is irreducible we must have

$f'(x) = 0$. Thus $f(x)$ is separable irreducible if and only if $f'(x) \neq 0$, so the question of whether a polynomial $p(x) \in F[x]$ is separable depends only on relationships within $F[x]$, and not on the choice of an extension E such that $p(x)$ has a factorisation into linear factors in $E[x]$.

We now turn our attention to finite fields. To begin we note that any finite subgroup of the multiplicative group of a field is cyclic. This is an exercise in group theory, showing that if a finite abelian group A satisfies the condition that any equation of the form $x^n = 1$ has at most n roots in A then A is cyclic. Given any field E we can examine the subfield F of E generated by 1. This is called the prime field of E and is either isomorphic to the rational field $\mathbb{Q}$ or to $\mathbb{Z}/p\mathbb{Z}$ for some prime p (if this is new it is a useful exercise). If E is a finite field then the prime field F has order p for some prime p and E has order p^n, where $n = [E : F] \geq 1$.

3.10 Theorem. For each prime p and natural number n there exists a field E of order p^n. If F is the prime field of order p E/F is a Galois extension, indeed a splitting field over F for the polynomial $x^{p^n} - x$, and $\mathrm{Gal}(E/F)$ is cyclic.

Proof. Let $F = \mathbb{Z}/p\mathbb{Z}$. Then F has order p. Let $f(x) = x^{p^n} - x \in F[x]$. Now f is a polynomial without repeated roots, since $f'(x) = -1 \neq 0$, so that if E is the splitting field for f over F then E is a Galois extension of F. We need to show that $[E : F] = n$.

Let S be the set of roots in E of $x^{p^n} - x$. We note that $|S| = p^n$. But if u and v are both in S then $(u + v)^{p^n} = u^{p^n} + v^{p^n} = u + v$ and $(uv)^{p^n} = uv$. Since S is finite these two statements imply S is a field, which clearly contains all the roots of $x^{p^n} - x$. Thus $S = E$ and $|E| = p^n$ as required.

Let α be the ring automorphism of E such that $u\alpha = u^p$, for $u \in E$. Since $u^p = u$ if $u \in F$ we see that $\alpha \in \mathrm{Gal}(E/F)$. Now $\alpha^r : u \to u^{p^r}$. Since the group of units of E is cyclic of order $p^n - 1$ there exists an element v of E such that $v^{p^n - 1} = 1$ but $v^r \neq 1$ for $0 < r < p^n - 1$.

Note that $v\alpha^{n-1} = v^{p^{n-1}} \neq v$ so α has order at least n. But $\alpha \in \mathrm{Gal}(E/F)$

and so α has order n and thus $Gal(E/F)$ is cyclic and generated by the automorphism $u \to u^p$. //

In the case $E \neq F$ the element $v \in E$ that was identified in the above proof has the property that $E = F(v)$, i.e. that no subfield K such that $F \subseteq K \subset E$ contains v. An element with this property is said to be a primitive element (for E over F). Suppose E, G are fields of p^n elements for some prime p and integer $n \geq 1$. Each of E and G is a splitting field for $x^{p^n} - x$ over the prime field $F = \mathbb{Z}/p\mathbb{Z}$, so by Theorem 3.2 $E \simeq G$. Thus we will often refer to "the field of order p^n". Finite fields are sometimes called Galois fields, and the field of order p^n is often denoted $GF(p^n)$, or $GF(q)$ where $q = p^n$ is a prime power. The automorphism $\alpha : x \to x^p$ of any Galois field $GF(p^n)$ is called the Frobenius automorphism.

RINGS OF ALGEBRAIC INTEGERS

In this section we are going to investigate the ring of integers, $\underline{O}$, of a finite extension K of the rational field $\mathbb{Q}$. Our main objective is to show that every non-zero prime ideal of $\underline{O}$ is maximal and that if $\underline{p} \neq 0$ is a prime ideal of $\underline{O}$ then $\underline{O}_{\underline{p}}$, the localisation at $\underline{p}$ is a commutative principal ideal domain with some other rather nice properties. We will show at the end of the section that such an $\underline{O}_{\underline{p}}$ is a noetherian domain with just one non-zero prime ideal, so $\underline{O}_{\underline{p}}$ is local.

A noetherian ring is one which satisfies the a. c. c. on ideals(recall in this chapter everything is commutative). This section begins by examining the structure of a domain satisfying the above hypothesis.

3.11 Lemma. Let R be a noetherian local domain with maximal ideal $\underline{m} \neq 0$ such that $\underline{m}$ is the only non-zero prime ideal of R. Then

(a) given any non-zero ideal $\underline{a}$ there exists a natural number n such that $\underline{a} \supseteq \underline{m}^n$, and

(b) $\underline{m}^n \neq \underline{m}^{n+1}$ for all $n \geq 0$.

Proof. (a) Let Σ be the set of non-zero ideals of R which do not contain a power of $\underline{m}$. If $\Sigma \neq \phi$ then Σ contains a maximal element, say $\underline{b}$, since R is noetherian. As $0 \neq \underline{b} \neq \underline{m}$, $\underline{b}$ is not prime. Thus there exists $x, y \in R$ such that $x, y \notin \underline{b}$ but $xy \in \underline{b}$. Now $xR + \underline{b}$ and $yR + \underline{b}$ both strictly contain $\underline{b}$. So each contains a power of $\underline{m}$. But $(xR + \underline{b})(yR + \underline{b}) \subseteq \underline{b}$. Then $\underline{b}$ would contain a power of $\underline{m}$ contradicting the definition of $\underline{b}$. So $\Sigma = \phi$.

(b) $\underline{m}^n \neq \underline{m}^{n+1}$ for all $n \geq 0$. Assume that for some $n \geq 0$ $\underline{m}^n = \underline{m}^{n+1}$. Since R is a domain and $\underline{m} \neq 0$, $\underline{m}^n \neq 0$. Since R is noetherian there is a maximal ideal $\underline{a} \subseteq \underline{m}^n$, such that $\underline{a} \neq \underline{m}^n$. So $\underline{m}^n/\underline{a}$ is a simple R-module. But $\underline{m}$ is the Jacobson radical of R so $\underline{m}$ annihilates $\underline{m}^n/\underline{a}$ (2.20). So $\underline{m}^n . \underline{m} \subseteq \underline{a}$, and $\underline{m}^{n+1} = \underline{m}^n$. Then $\underline{a} = \underline{m}^n$ which is a contradiction. Hence $\underline{m}^n \neq \underline{m}^{n+1}$ for all $n \geq 0$. //

3.12 Proposition. Let R be a noetherian local domain with maximal ideal $\underline{m} \neq 0$ such that $\underline{m}$ is the only non-zero prime ideal of R.

Then the following are equivalent:

(i) R is integrally closed in its field of fractions, K;

(ii) $\underline{m}$ is a principal ideal;

(iii) every non-zero ideal is a power of $\underline{m}$

(iv) there exists $x \in R$ such that every non-zero ideal of R is of the form $x^k R$ for some $k \geq 0$.

Proof. (i) $\Rightarrow$ (ii). Let $a \in \underline{m}$ with $a \neq 0$. By 3.11(a) there is a power of $\underline{m}$, say $\underline{m}^n$ such that $Ra \supseteq \underline{m}^n$ and $Ra \not\supseteq \underline{m}^{n-1}$. Choose $b \in \underline{m}^{n-1}$ with $b \notin Ra$. Let $x = a/b$ where $0 \neq x \in K$ the field of fractions of R. Now if $x^{-1} \in R$, then $x^{-1}.a = b \in Ra$ so $x^{-1} \notin R$ and so x^{-1} is not integral over R. If $x^{-1}\,\underline{m} \subseteq \underline{m}$ then $\underline{m}$ would be a faithful $R[x^{-1}]$-module. But then by (2.5) x^{-1} would be integral over R. Hence $x^{-1}\underline{m} \not\subseteq \underline{m}$. But $b\underline{m} \subseteq Ra$ so $(b/a)\,\underline{m} \subseteq R$, i.e. $x^{-1}\underline{m} \subseteq R$. Thus $x^{-1}\underline{m} = R$ or $\underline{m} = Rx$.

Since $\underline{m} \neq \underline{m}^2$ there exists $y \in \underline{m}$ such that $y \notin \underline{m}^2$. So $y = rx$ with $r \notin \underline{m}$. Then r is a unit of R, so $x = \frac{1}{r} . y$ and $\underline{m} = Ry$ and $\underline{m}$ is a principal ideal.

(ii) ⇒ (iii). Let $\underline{a}$ be any non-zero ideal of R. If $\underline{a} = R$ then $\underline{a} = \underline{m}^0$, as required. If $\underline{a} \neq R$ then $\underline{a} \subseteq \underline{m}$ so by (3.11(a)) above there exists a least integer t such that $\underline{m}^t \subseteq \underline{a}$. If $\underline{m}^t \neq \underline{a}$ there exists an integer j, $1 \leq j < t$, and an element $z \in R$ such that $zy^j \in \underline{a}$, $zy^j \notin \underline{m}^t$ and $z \notin \underline{m}$. But then z is a unit of R, so $z^{-1}zy^j = y^j \in \underline{a}$, that is $\underline{m}^j \subseteq \underline{a}$, contradicting the minimality of t. Hence $\underline{a} = \underline{m}^t$.

(iii) ⇒ (iv). Choose $x \in \underline{m}$ with $x \notin \underline{m}^2$ (using (3.11(b))). Then $\underline{m} = Rx$ as before and so by (iii) if $\underline{a}$ is an ideal of R $\underline{a} = \underline{m}^t = Rx^t$.

(iv) ⇒ (i). Observe that $Rx = \underline{m}$ and that $Rx^t \neq Rx^{t+1}$ for all t. Given any element r in R, there exists a unique n such that $r \in \underline{m}^n$ but $r \notin \underline{m}^{n+1}$. Then $r = ux^n$ where u is a unit in R.

Let $y \in K$ and let $y = r/s$, with $r, s \in R$. Assume $r = u_1x^{n_1}$ and $s = u_2x^{n_2}$ where u_1 and u_2 are units. Note that either r/s or s/r is in R depending on whether $n_1 \geq n_2$ or $n_2 \geq n_1$. Thus either $y \in R$ or $\frac{1}{y} \in R$. Assume y is integral over R. If $y \in R$ there is nothing to prove. Let $y \notin R$ and $y^m + a_1y^{m-1} + \ldots + a_m = 0$, for $a_1, \ldots, a_m \in R$. As $y \notin R$, $\frac{1}{y} \in R$ and so $y = -(a_1 + a_2/y + a_3/y^2 + \ldots + a_m/y^{m-1}) \in R$. So R is integrally closed in K. //

Notice that under these hypotheses R is a principal ideal domain and every non-zero ideal of R is of the form $\underline{m}^n$.

We now have to show that the localisation of the ring of integers of a finite extension of $\mathbb{Q}$ satisfies these hypotheses. So for the remainder of this section we assume $\mathbb{Z} \subseteq \mathbb{Q} \subseteq K \subset \mathbb{C}$ where K is a finite extension of $\mathbb{Q}$ and we let $\underline{O}$ be the ring of integers of K over $\mathbb{Z}$.

<u>3.13 Lemma.</u> Every non-zero prime ideal of $\underline{O}$ is maximal.

<u>Proof.</u> Let $P \neq 0$ be a prime ideal of $\underline{O}$. We now show that $P \cap \mathbb{Z} \neq 0$. Let $a \in P$ and choose $p(x) \in \mathbb{Z}[x]$ such that p(x) is monic and of minimal degree for $p(a) = 0$, so if $p(x) = r_0 + r_1x + \ldots + x^r$, then $r_0 \neq 0$. Clearly $r_0 = -a(r_1 + \ldots + a^{n-1}) \in P \cap \mathbb{Z}$. So $P \cap \mathbb{Z}$ is a non-zero prime ideal of $\mathbb{Z}$ and hence maximal. By (2.9) it follows that P is maximal. //

Before beginning the proof of the next theorem we need to make some comments concerning linear algebra. Let K be a finite extension of $\mathbb{Q}$. If $x \in K$ then x defines a $\mathbb{Q}$-linear transformation $K \to K$ by $a \to ax$, $a \in K$. Using this linear transformation x has a characteristic polynomial and also a trace function $\mathrm{Tr}(x)$. Recall that $\mathrm{Tr}(x)$ is a term of the characteristic polynomial.

Since K is a finite extension of $\mathbb{Q}$ x is algebraic over $\mathbb{Q}$ and then x has a minimum polynomial which is irreducible. However x has a minimum polynomial as a linear transformation but it is not difficult to show that these two minimum polynomials are identical. Further since the characteristic polynomial of a linear transformation has the same irreducible factors as the minimum polynomial we have that the characteristic polynomial for x is a power of its minimum polynomial since the latter is irreducible. As a consequence $\mathrm{Tr}(x)$ is a function of the coefficients of the minimum polynomial of x.

If we happened to have chosen x to be an integer in K then x satisfies a polynomial over $\mathbb{Z}$. It is a straightforward exercise (using Gauss' Lemma) to show that the minimum polynomial has coefficients in $\mathbb{Z}$. Alternatively we can consider a splitting field for the minimum polynomial and observe that since all the roots are conjugate under the action of the Galois group they are all algebraic integers and hence so are the coefficients of the minimum polynomial. Then $\mathrm{Tr}(x) \in \mathbb{Z}$ if x is an integer in K.

Also we observe that $\mathrm{Tr}(xy)$ is a symmetric bilinear form on K, that is $\mathrm{Tr}(xy)$ is linear in both the first factor and the second factor. Finally this form is non-degenerate since $\mathrm{Tr}(xx^{-1}) = [K : \mathbb{Q}] \neq 0$ for $x \neq 0$.

<u>3.14 Theorem.</u> $\underline{O}$ is a noetherian ring.

<u>Proof.</u> Suppose that $b \in K$. Since K is a finite extension of $\mathbb{Q}$, b is algebraic over $\mathbb{Q}$, with minimum polynomial say $x^n + a_1 x^{n-1} + \dots + a_n \in \mathbb{Q}[x]$. We can put $a_1, \dots, a_n$ over a common denominator, say $a_i = c_i/d$ for each i, where $0 \neq d$ and $c_i, d \in \mathbb{Z}$. It then follows easily that

$$(db)^n + c_1 (db)^{n-1} + \dots + c_n d^{n-1} = 0 ,$$

so bd is integral over $\mathbb{Z}$. Thus if $b_1, \ldots, b_n$ is a basis of K over $\mathbb{Q}$ we may assume (if necessary by multiplying by suitable non-zero elements of $\mathbb{Z}$) that $b_1, \ldots, b_n$ are integral elements of K.

We use the symmetric bilinear form $\mathrm{Tr}(xy)$ defined in the comments preceeding the statement. Then there exists a dual basis $b'_1, \ldots, b'_n$ such that $\mathrm{Tr}(b_i b'_j) = \delta_{ij}$, the Kronecker δ.

Let $x \in \underline{O}$ and let

$$x = \sum_{i=1}^{n} b'_i x_i \qquad \text{where } x_i \in \mathbb{Q}.$$

Since $x \in \underline{O}$ and $b_i \in \underline{O}$, $b_i x \in \underline{O}$ for $1 \le i \le n$. By the comments preceeding the theorem, $\mathrm{Tr}(b_i x) \in \mathbb{Z}$. So $\mathrm{Tr}(b_i x) = \Sigma\,\mathrm{Tr}(b_i b'_j x_j) = \Sigma x_j \delta_{ij} = x_i$. Thus $x_i \in \mathbb{Z}$, $1 \le i \le n$. Hence $\underline{O}$ is a submodule of the free $\mathbb{Z}$-module $\Sigma\,\mathbb{Z}b_i$. As $\mathbb{Z}$ is a principal ideal domain the result now follows from results about modules over principal ideal domains (or more specifically finitely generated abelian groups). //

Now we can choose P to be a prime ideal of $\underline{O}$. Then $\underline{O}_P$ is a local domain by the results proved in Chapter 2. Also observe that $\underline{O}$ is integrally closed in K by its definition and so by (2.10) $\underline{O}_P$ is integrally closed in K. The remaining properties of $\underline{O}_P$, i.e. that it is Noetherian and that all its non-zero prime ideals are maximal, follow from (3.13) and (3.14) and the results proved about ideals and localisations in Chapter 2.

Finally if p is a prime in $\mathbb{Z}$ by (2.11) there is a prime ideal P in $\underline{O}$ such that $P \cap \mathbb{Z} = p\mathbb{Z}$. Then $\underline{O}_P/\underline{m}$ is a field of characteristic p where $\underline{m}$ is the unique maximal ideal of $\underline{O}_P$.

As an example it is useful to consider $K = \mathbb{Q}(\sqrt{d})$ where d is a square free natural number. The question is, what does $\underline{O}$ look like in this case? It is fairly clear that $\underline{O} \supseteq \mathbb{Z}[\sqrt{d}]$, since $(\sqrt{d})^2 - d = 0$ so $(\sqrt{d}) \in \underline{O}$.

It is possible to show that $\underline{O} = \mathbb{Z}[\sqrt{d}]$ if $d \not\equiv 1 \pmod 4$ but that $\underline{O} = \mathbb{Z}[(\tfrac{1}{2} + \tfrac{1}{2}\sqrt{d})]$ if $d \equiv 1 \pmod 4$. Thus $\tfrac{1}{2} + \frac{\sqrt{5}}{2}$ is an integer in $\mathbb{Q}(\sqrt{5})$.

In $\mathbb{Q}(\sqrt{3})$ the ring of integers $\underline{O}$ is precisely $\mathbb{Z}[\sqrt{3}]$, which can be written as $\mathbb{Z} + \sqrt{3}\mathbb{Z}$ as $\mathbb{Z}$-module. In $\underline{O}$, 2 has the factorisation $2 = (\sqrt{3}+1)(\sqrt{3}-1)$, and it is easy to check that neither factor is a unit of $\underline{O}$. Furthermore, a routine check shows that $(\sqrt{3}+1)(2-\sqrt{3}) = \sqrt{3}-1$ and $(\sqrt{3}-1)(2+\sqrt{3}) = \sqrt{3}+1$, so the ideals of $\underline{O}$ generated by the two factors are in fact the same ideal, say A. So we have $2\underline{O} = A^2$. It is a worthwhile but not difficult exercise to show that A is the only prime ideal of $\underline{O}$ which contains 2, and that $\underline{O}/A$ is a field of order 2. In the same ring the rational prime 13 exhibits slightly different behaviour. We have the factorisation $13 = (4+\sqrt{3})(4-\sqrt{3})$ and it is an easy exercise to check that $4+\sqrt{3}$ and $4-\sqrt{3}$ generate distinct maximal ideals P and Q of $\underline{O}$, so that $13\underline{O} = PQ$ and $\underline{O}/P$ and $\underline{O}/Q$ are (isomorphic) fields of order 13. It is a significantly more difficult task to show that the rational prime 7 has no non-trivial factors in $\underline{O}$, so that $\underline{O}/7\underline{O}$ is a field of $7^2 = 49$ elements. Note that, in our discussion of the factorisation of 2 we implicitly proved that $2 \pm \sqrt{3}$ are units of $\underline{O}$; in fact $(2+\sqrt{3})(2-\sqrt{3}) = 1$.

Although the example just given is a principal ideal domain, some rings of algebraic integers are not principal ideal domains. That is why the process of localising at a prime, which does yield a principal ideal domain, is important in the applications which follow.

It is also worth looking at what happens when we localise at a prime ideal of a ring of algebraic integers. In the case of $\mathbb{Z}$, the local ring at the ideal generated by the rational prime p is denoted by $\mathbb{Z}_p$ (rather than $\mathbb{Z}_{(p)}$) and $\mathbb{Z}_p = \{q/r : q, z \in \mathbb{Z} \text{ and } p \nmid r\}$. Thus $\mathbb{Z}_p \subseteq \mathbb{Q}$, and $\mathbb{Z}_p$ comprises the rationals whose denominator is not divisible by p. In the example above, if A is the prime ideal $(1+\sqrt{3})\underline{O}$ of $\underline{O} = \mathbb{Z}[\sqrt{3}]$, then it can be shown that $A = \{a + b\sqrt{3} : a, b \in \mathbb{Z}, a+b \text{ is even}\}$. Since $\underline{O}$ is an integral domain and $\sqrt{3} \in \mathbb{R}$ we clearly have $\underline{O}_A \subseteq \mathbb{R}$, and then

$$\begin{aligned}\underline{O}_A &= \{x/y : x, y \in \underline{O}, y \notin A\} \\ &= \{(a+b\sqrt{3})/(c+d\sqrt{3}) : a, b, c, d \in \mathbb{Z}, c+d \text{ is odd}\}.\end{aligned}$$

CYCLOTOMIC POLYNOMIALS AND PRIMITIVE PRIME DIVISORS

In this section we introduce the cyclotomic polynomials $\Phi_n(x)$ and discuss the divisibility of $\Phi_n(q)$ where q is a rational integer. But first we need some preliminaries concerning arithmetic functions.

Let f be a function from $\mathbb{N} \to \mathbb{N}$. Then f is said to be multiplicative if $f(ab) = f(a)f(b)$ whenever a, b are coprime, i.e. $(a, b) = 1$. Clearly the identity function $n \to n$ is multiplicative. We shall need to look at two other functions on $\mathbb{N}$ which will play an important role. The first is the Mobius function $\mu(n)$. This is defined in terms of a prime decomposition of n. If $n = p_i^{e_i} \dots p_r^{e_r}$ where each $e_i > 0$ and $\{p_1, \dots, p_r\}$ is a set of r distinct primes then by definition

$$\mu(n) = \begin{cases} 1 & \text{if } n = 1 \\ (-1)^r & \text{if } e_1 = e_2 \dots = e_r = 1 \\ 0 & \text{otherwise.} \end{cases}$$

It is easy to check that μ is multiplicative.

The second function is the Euler 'phi' function $\phi(n)$. This is defined to be the number of integers between 1 and n - 1 inclusive which are coprime to n. Alternatively it is the order of the group of units of the ring $\mathbb{Z}/n\mathbb{Z}$ or again the order of the automorphism group of the cyclic group of order n. By using whichever of these definitions which most appeals it is not too difficult to check that ϕ is a multiplicative function.

The Mobius function also satisfies the equation

$$\sum_{d|n} \mu(d) = \begin{cases} 1 & \text{if } n = 1 \\ 0 & \text{if } n > 1. \end{cases}$$

The Mobius transform $g(n)$ of a function $f : \mathbb{N} \to \mathbb{N}$ is defined by

$$g(n) = \sum_{d|n} \mu(\tfrac{n}{d}) f(d) = \sum_{d|n} \mu(d) f(\tfrac{n}{d}) .$$

Then

$$\sum_{t|n} g(t) = \sum_{t|n} \left(\sum_{d|t} \mu\left(\frac{t}{d}\right) f(d) \right)$$

$$= \sum_{d|t} \left(\sum_{t|n} \mu\left(\frac{t}{d}\right) f(d) \right)$$

$$= \sum_{d|t} f(d) \left(\sum_{\substack{t|n \\ d|t}} \mu\left(\frac{t}{d}\right) \right) .$$

Note that the inner sum is over all divisors of n which are multiples of d. So the inner sum is equal to $\sum_{x|(n/d)} \mu(d)$, which is non-zero if and only if $n = d$.

So

$$\sum_{t|n} g(t) = f(n) .$$

This is called the Mobius inversion formula.

Since ϕ is multiplicative, when $n = p_1^{e_1} \ldots p_r^{e_r}$ is a prime decomposition of n we have $\phi(n) = \phi(p_1^{e_1}) \ldots \phi(p_r^{e_r})$. However $\phi(p_i^{e_i})$ is easy to compute and is $p^{e_1} - p^{e_1 - 1}$.

So $\phi(n) = n \prod_{i=1}^{r} (1 - \frac{1}{p_i})$ and hence

$$\phi(n) = n \sum_{d|n} \mu(d)/d$$

$$= \sum_{d|n} \mu(d) \frac{n}{d} = \sum_{D|n} \mu\left(\frac{n}{D}\right) D$$

Thus ϕ is the Mobius transform of the identity and by the Mobius inversion formula

$$n = \sum_{t|n} \phi(t) .$$

We now turn our attention to n^{th} roots of unity. Let F be a field and assume that the characteristic of F is either zero or prime to n. We will fix this assumption for the next few pages. Let $p(x) = x^n - 1$. Then $p'(x) = nx^{n-1}$ which is non-zero and hence clearly coprime to $x^n - 1$. Thus there is a splitting field E for p(x) over F which is a Galois extension of F. A root of

$x^n - 1$ in E will be called an nth root unity. The precise reference to the field E will often be omitted, but note that there are n such roots.

A primitive n^{th} root of unity is a root ξ such that $\xi^n = 1$ but $\xi^m \neq 1$ for $0 < m < n$.

3.15 Lemma. Under the assumptions above there are $\phi(n)$ primitive n^{th} roots of unity.

Proof. We prove the result by induction on n. If $n = 1$ then clearly 1 is the only (first) root of unity and is primitive. Assume the result is true for all natural numbers less than n. Let ξ be any n^{th} root of unity. Then ξ is a primitive d^{th} root of unity for some d. Now $d|n$ as the nth roots of unity form a group of order n, and so there are $\phi(d)$ primitive d^{th} roots of unity, if $d < n$. We can count all the nth roots of unity which are not primitive n^{th} roots, by

$$\sum_{\substack{d|n \\ d<n}} \phi(d).$$

But there are n roots altogether and so the number of nth roots is

$$n - \sum_{\substack{d|n \\ d<n}} \phi(n) = \phi(n), \qquad \text{since}$$

$$n = \sum_{d|n} \phi(d) \ . \ //$$

Given n we define the cyclotomic polynomial $\Phi_n(x) = \Pi(x - \xi)$ where ξ runs through the set of primitive n^{th} roots of unity, in some splitting field F for $x^n - 1$ over E. In fact $\Phi_n(x) \in E[x]$ does not depend on F. For $\Phi_n(x)$ has degree $\phi(n)$ as a consequence of (3.15), and also we have

$$x^n - 1 = \prod_{d|n} \Phi_d(x) \ .$$

Hence by a multiplicative analogue of the Mobius inversion formula, we obtain an expression for $\Phi_n(x)$,

$$\Phi_n(x) = \prod_{d|n} (x^d - 1)^{\mu(n/d)}$$

where every term $(x^d - 1)$ is in $E[x]$. This enables the $\Phi_n(x)$ to be calculated explicitly. For example

$$\Phi_6(x) = (x-1)^{\mu(6)} \cdot (x^2-1)^{\mu(3)} (x^3-1)^{\mu(2)} (x^6-1)^{\mu(1)}$$

$$= \frac{(x-1)(x^6-1)}{(x^2-1)(x^3-1)} = \frac{x^3+1}{x+1} = x^2 - x + 1 .$$

<u>3.16 Lemma.</u> The cyclotomic polynomials over $\mathbb{Q}$ have integer coefficients.

<u>Proof.</u> We observe that since n^{th} roots of unity are algebraic integers the coefficients of the cyclotomic polynomials are algebraic integers. We use induction on n. Notice $\Phi_1(x) = x - 1$ so $\Phi_1(x) \in \mathbb{Z}[x]$.

Since $x^n - 1 = \Phi_n(x) \left(\prod_{\substack{d|n \\ d<n}} \Phi_d(x) \right)$ we see that $x^n - 1 = \Phi_n(x)\, q(x)$ where $q(x) \in \mathbb{Z}[x]$. By solving the equations for the coefficients of $\Phi_n(x)$ in terms of $q(x)$ and $x^n - 1$ we have that the coefficients of $\Phi_n(x)$ are rational numbers. But they are also algebraic integers, so they are rational integers as required. //

<u>3.17 Lemma.</u> If $q \in \mathbb{N}$ and $n \in \mathbb{N}$ with $q, n > 1$ then $\Phi_n(q) \in \mathbb{N}$.

<u>Proof.</u> This follows immediately from 3.16 and the formula $\Phi_n(x) = \prod_{d|n} (x^d - 1)^{\mu(n/d)}$. //

Now $x^n - 1 = \prod_{d|n} \Phi_d(x)$, so if $t|n$ then $x^n - 1 = \Phi_n(x)\, q(x) \prod_{d|t} \Phi_d(x)$ for some $q(x) \in \mathbb{Z}[x]$. Thus $x^n - 1 = \Phi_n(x)\, q(x)\, (x^t - 1)$ and so $\Phi_n(x) \,|\, ((x^n - 1)/(x^t - 1))$ if $t|n$. //

We now concentrate on cyclotomic polynomials over $\mathbb{Q}$ and assume we are working in $\mathbb{Z}[x]$.

<u>3.18 Lemma.</u> Let $n \in \mathbb{N}$ and $n = pm$ where p is prime and $(p, m) = 1$. Then

$\Phi_m(x^p) = \Phi_n(x)\Phi_m(x)$.

Proof. $\Phi_n(x) = \prod_{d|mp} (x^d - 1)^{\mu(n/d)}$

$$= \prod_{d|m} (x^d - 1)^{\mu(mp/d)} \cdot \prod_{d|m} (x^{pd} - 1)^{\mu(m/d)} .$$

But $\mu(mp/d) = -\mu(m/d)$ wherever $d|m$, since p is prime. So $\Phi_n(x) = (\Phi_m(x))^{-1}\Phi_m(x^p)$ as claimed. //

3.19 Lemma. Let $n = p^{a+1}m$ where p is a prime, $m \in \mathbb{N}$ and $(m, p) = 1$ and $a \in \mathbb{N}$, $a \geq 1$. Then $\Phi_n(x) = \Phi_{pm}(x^{p^a})$.

Proof. Again we write

$$\Phi_n(x) = \prod_{d|n} (x^d - 1)^{\mu(n/d)} .$$

Since $\mu(r) = 0$ if $p^2 | r$ we obtain

$$\Phi_n(x) = \prod_{d|pm} (x^{p^a d} - 1)^{\mu(pm/d)}$$

$$= \phi_{pm}(x^{p^a}) . \text{ //}$$

We need one last simple lemma before embarking on the main results in this section.

3.20 Lemma. If $q, f \in \mathbb{N}$ and p is a prime such that $q^f \equiv 1 \pmod p$ then $q^{pf} \equiv 1 \pmod{p^2}$.

Proof. Let $q^f = 1 + cp$, $c \in \mathbb{N}$. Then $q^{fp} = (1 + cp)^p = 1 + cp^2 + \ldots \equiv 1 \pmod{p^2}$. //

3.21 Theorem. Let q and n be natural numbers both larger than 1. Assume that p is a prime and that p divides $\Phi_n(q)$ and let q have order $f \pmod p$.

Then $n = fp^i$ for some $i \in \mathbb{Z}$, $i \geq 0$. If $i > 0$ then p is the largest prime divisor of n. If $i > 0$ and p^2 divides $\Phi_n(q)$ then $n = p = 2$.

Proof. Since $p | \Phi_n(q)$ and $\Phi_n(q) | (q^n - 1)$ it follows that $f | n$. Put $n = fp^i w$ where $i \geq 0$, and $(p, w) = 1$. Assume $w > 1$ and let $r = fp^i$. Thus $q^r \equiv 1 \pmod p$ and since $r < n$, $\Phi_n(q)$ divides $(q^n - 1)/(q^r - 1)$. But

$$\frac{q^n - 1}{q^r - 1} = (q^r)^{w-1} + (q^r)^{w-2} + \dots + q^r + 1 \equiv w \pmod p.$$

Thus $p | \Phi_n(q) \,|\, \dfrac{(q^n - 1)}{(q^r - 1)}$ which contradicts the assumption that $(w, p) = 1$ and $r < n$. Thus $r = n$ and $w = 1$.

So $n = fp^i$ and since f is the order of $q \pmod p$, $f | (p-1)$ and so $f < p$ (and $(f, p) = 1$).

In the case $i > 1$ we use a similar argument to show that if $s = fp^{i-1}$, then $\Phi_n(q) | \{(q^n - 1)/(q^s - 1)\}$. But

$$\frac{q^n - 1}{q^s - 1} = (q^s)^{p-1} + \dots + q^s + 1 ,$$

and $q^s = q^{fp^{i-1}} \equiv 1 \pmod{p^2}$ by (3.20). Thus $\left\{\dfrac{(q^n - 1)}{(q^s - 1)}\right\} \equiv p \pmod{p^2}$. So if $i > 1$ then $p^2 \nmid \Phi_n(q)$.

Assume $p^2 | \Phi_n(q)$ and so $n = f$ or $n = fp$. Assume $n = fp$. We now obtain the equation

$$\frac{(q^n - 1)}{(q^f - 1)} = (q^f)^{p-1} + (q^f)^{p-2} + \dots + q^f + 1.$$

Let $q^f = 1 + cp$, so that

$$\frac{(q^n - 1)}{(q^f - 1)} \equiv 1 + (p - 1)cp + 1 + (p - 2)cp + \dots + 1 + cp + 1 \pmod{p^2}.$$

So

$$\frac{(q^n - 1)}{(q^f - 1)} \equiv p + cp\left(\frac{(p - 1)p}{2}\right) \pmod{p^2}.$$

If $p \neq 2$ then $\dfrac{(p - 1)p}{2} \equiv 0 \pmod p$ and hence $(q^n - 1)/(q^f - 1) \equiv p \pmod{p^2}$. Thus if $p^2 | \Phi_n(q)$ and $i > 0$, we must have $p = 2$.

In that case, $f = 1$ and $n = 2^i$. But when n is even $q^n - 1 = (q^{n/2}-1)(q^{n/2}+1)$ and so $q^{n/2} + 1 = (q^n - 1)/(q^{n/2} - 1)$ and $\Phi_n(q) \mid (q^{n/2} + 1)$. So if $n = 2^i$ and $i > 1$, it follows that $q^{n/2} + 1 = q^{2^{i-1}} + 1 \equiv 2 \pmod 4$, which is a contradiction. Thus $i = 1$ and $n = 2$. //

<u>3.22 Corollary.</u> Let s be the highest common factor of n and $\Phi_n(q)$. Either $s = 1$ or s is prime.

<u>Proof.</u> This is immediate from (3.21). //

<u>3.23 Theorem.</u> Let q and n be natural numbers both greater than 1 and let p be the highest common factor of $\Phi_n(q)$ and n. Let p^j be the largest power of p which divides $\Phi_n(q)$ (so $p^j = 1$ if $p = 1$), and let $\Phi^*_n(q) = \Phi_n(q)p^{-j}$.

Then $\Phi^*_n(q) = 1$ if and only if either $n = 2$ and $q + 1$ is a power of 2 or $n = 6$ and $q = 2$.

<u>Proof.</u> Firstly assume $n = 2$. Then $\Phi_2(q) = q + 1$ and $\Phi^*_2(q) = 1$ if and only if $q + 1$ is a power of 2. Similarly note that $\Phi_6(x) = x^2 - x + 1$ and $\Phi_6(2) = 4 - 2 + 1 = 3$. Then $3\Phi^*_6(2) = \Phi_6(2)$ and $\Phi^*_6(2) = 1$. Thus the exceptions do occur.

Let $n \geq 3$ and assume $\Phi^*_n(q) = 1$. So $p \neq 1$ and by (3.22) p is a prime. By (3.21) if $p \mid n$ and $p^2 \mid \Phi_n(q)$ then $n = p = 2$. So since $n \geq 3$ and p does divide n, $p^2 \nmid \Phi_n(q)$. So $\Phi_n(q) = p$. But $\Phi_n(x) = \Pi(x - \xi)$ where ξ runs through the primitive complex n^{th} roots of unity so

$$|\Phi_n(q)| = |\Pi(q - \xi)| = \Pi|q - \xi| > (q - 1)^{\phi(n)},$$

since $|q - \xi| \geq q - 1$ for all ξ and there are $\phi(n)$ distinct factors.

But $n = fp^i$ where f is the order of q by (3.21) and since $p \mid n$, $i > 0$. Since $p \mid (q - 1)$, p and q are unequal. If $q \neq 2$ then $p > (q - 1)^{(p-1)}$, which is impossible. Hence $q = 2 \neq p$.

Suppose it were the case that $p^2 \mid n$. From (3.19) we would have $p = \Phi_n(2) = \Phi_{pf}(2^{p^{j-1}})$ for some $j > 1$, so as above $p > (2^{p^{j-1}} - 1)^{\phi(pf)} \geq$

$(2^{p^{j-1}} - 1)^{p-1}$, which is impossible. Hence $p^2 \nmid n$, so $n = pf$. From (3.18)

$$\Phi_f(2^p) = \Phi_n(2)\Phi_f(2) .$$

But $\Phi_{fp}(2) = \Phi_n(2) = p$, and $\Phi_f(2)$ divides $2^f - 1$, and $f|(p-1)$, so

$$p(2^{p-1} - 1) \geq \Phi_{fp}(2)\Phi_f(2) = \Phi_f(2^p) \geq (2^p - 1)^{\phi(f)} .$$

Hence $p \geq (2^p - 1)^{\phi(f) - 1}$, so $\phi(f) = 1$, and $f = 1$ or 2. Since f is the order of 2 mod p, and $p \neq 2$, we must have $f = 2$, and $4 \equiv 1$ mod p. That is, $f = 2$, $p = 3$ and $n = 6$, as claimed. //

If q, n are integers greater than 1, a <u>primitive divisor</u> of $q^n - 1$ is a natural number t such that $t|(q^n - 1)$ and $(t, (q^d - 1)) = 1$ for $1 \leq d < n$. A primitive divisor p of $q^n - 1$ which is also prime is called a <u>primitive prime divisor</u> of $q^n - 1$, and is characterised by the properties that $p|(q^n - 1)$ but $p \nmid (q^d - 1)$ for $1 \leq d < n$. The preceding set of results will now be employed to prove a key theorem with a long pedigree going back to Birkhoff and Vandiver [1] and **Zsigmondy** [22].

<u>3.24 Theorem.</u> Let q and n be natural numbers, both distinct from 1. Either there exists a primitive prime divisor of $q^n - 1$, or $n = 2$ and $q + 1$ is a power of 2, or $n = 6$ and $q = 2$.

<u>Proof.</u> Let p be a prime which divides $q^n - 1$, and let $d = d(p)$ be the order of q mod p. Clearly $d|n$, and if $d < n$ then

$$\frac{(q^n - 1)}{(q^d - 1)} \equiv (n/d) \bmod p .$$

Assume that $q^n - 1$ has no primitive prime divisor, so that by definition $d = d(r) < n$ for every prime r dividing $q^n - 1$. Let p be a prime dividing $\Phi_n(q)$. From above, if $d = d(p)$ then $(n/d) \equiv 0$ mod p, so by (3.22) $p = \mathrm{hcf}(n, \Phi_n(q))$. Furthermore, $\Phi_n(q)$ can have no other prime divisors, so $\Phi^*_n(q) = 1$. The result then follows from (3.23).//

In closing this chapter, we shall note three final results. The first is very straightforward.

3.25 Proposition. Let q and n be natural numbers both greater then 1, and let $\Phi^*_n(q)$ be defined as in (3.23). Then $(\Phi^*_n(q), n) = 1$.

Proof. This follows at once from (3.22) and (3.23), which show that h. c. f. $(\Phi_n(q), n)$ is either 1 or a prime which does not divide $\Phi^*_n(q)$. //

Secondly, we look at a special situation in which n is a power of 2, and $\Phi^*_n(q) = r = n + 1$ for some prime r.

3.26 Proposition. Suppose n, a, r are natural numbers such that $n = 2^a$ and $n + 1 = r$ is prime. If q is an odd natural number such that $\Phi^*_n(q) = r$ then either $n = 2$ or $n = 4$ and $q = 3$.

Proof. By (3.19) we have $\Phi_{2^a}(x) = \Phi_2(x^{2^{a-1}}) = x^{2^{a-1}} + 1$, so $\Phi_{2^a}(q) = q^{2^{a-1}} + 1$. Since $\Phi^*_n(q) = r = 2^a + 1$, $\Phi_n(q) \neq \Phi^*_n(q)$. Since 2 divides both n and $\Phi_n(q)$, we must from (3.21) have $\Phi_n(q) = r2^b$ for some $b \geq 1$. That is, $q^{2^{a-1}} + 1 = r2^b$. If $a > 1$ then $q^{2^{a-1}} \equiv 1 \bmod 4$, so $r2^b \equiv 2 \bmod 4$, which implies $b = 1$, and hence $q^{2^{a-1}} + 1 = (2^a + 1)2$. Since $a > 1$ we must have $q \geq 3$, and then $(q^{2^{a-1}} + 1)/(2^a + 1)$ is a monotone increasing function of q, which is easily seen by inspection to take the value 2 only when $q = 3$ and $a = 2$. //

Last we use this result to prove Wedderburn's theorem about finite division rings.

3.27 Theorem. Let K be a finite division ring. Then K is a field.

Proof. Let Z be the centre of K. It is straightforward to check that Z is a field and so $|Z| = q$ where q is a prime power. Also if K has dimension n over Z then K has order q^n. If $n = 1$ then $K = Z$ and the theorem is true.

So suppose $n > 1$, and for $y \in K$ let $C_K(y) = \{z \in K : zx = xz\}$.

If x is any element of K then $C_K(x)$ is a subspace of K. Thus if $x \notin Z$, $|C_K(x)| = q^d$ for $d < n$. Using the conjugacy class equation for the multiplicative group of K we obtain the following equation

$$q^n - 1 = |Z^{\#}| + \sum_{\substack{d<n \\ d \mid n}}{}' \frac{q^n - 1}{q^d - 1},$$

where the ' signifies that the sum is not necessarily over all possible divisions d (and where $Z^{\#}$ is the multiplicative group of Z).

As $|Z^{\#}| = q - 1$, if p is a primitive prime divisor of $q^n - 1$ we obtain a contradiction since p divides $(q^n-1)/(q^d-1)$ for $d < n$ and $d \mid n$.

This gives a contradiction unless $n = 2$ or $n = 6$ and $q = 2$ (3.24). If $n = 2$ the equation becomes

$$q^2 - 1 = (q - 1) + e(q + 1),$$

for some integer e. Then $q \mid e$ and $e(q + 1) + (q - 1) > q^2 - 1$ so there is a contradiction in this case. We leave the last case as an exercise for the reader. //

4 Representation theory

INTRODUCTION

In this chapter we develop ways of using the theory of non-commutative rings to obtain information about groups. The basic idea is that, for a given (commutative) field K and natural number n, the group GL(n, K) of automorphisms of an n-dimensional K-vector space is reasonably well-behaved and well understood. Thus by using homomorphisms of an arbitrary group H into groups GL(n, K), for suitable choices of K and n, we may obtain information about H itself.

It is worth remarking that GL(n, K) is well-behaved and well-understood for a particular reason: if V is an n-dimensional K-vector space, if S^V is the group of permutations of V and E is the ring of K-endomorphisms then

$$GL(n, K) = U(E) = S^V \cap E;$$

the elements of GL(n, K) are exactly the elements of S^V which are also elements of E. Hence a GL(n, K) is a permutation group with "well-behaved" elements.

The study of the non-commutative rings involved in this type of representation is one of the oldest parts of non-commutative ring theory, dating back at least as far as the work of Frobenius in the late nineteenth century. But it is a paradox of its early development, in the context of group representations, that much of the terminology and many of the ideas in the area are slightly tangential to the mainstream of modern ring theory. Hence it is sometimes hard for the ring theorist or the representation theorist to "translate" results about rings into results about representations, or vice versa. In what follows, we have tried to draw attention to these two different ways of looking at the results we develop, and to consistently provide a "translation" in the hope of an enriched understanding of both approaches.

ALGEBRAS

The basic concept in this chapter is that of an algebra. Let C be a commutative ring; then a C-algebra is a ringlet A such that A is also a left C-module, and that

$$c(xy) = (cx)y = x(cy)$$

for any elements x, y of A and c of C. The requirement that A be a left C-module might appear arbitrary, but we recall from Chapter 2 that, in the case of a one-sided module M over a commutative ring C, it is straightforward to endow M with a C - C bimodule structure according to the rule cm = mc for every c in C and m in M. If we adopt this approach to a C-algebra A as just defined, then A also acquires a right C-module structure, and it is easy to deduce that

$$(xy)c = x(yc) = (xc)y$$

for any elements x, y of A and c of C. Conversely, if we had used these rules and a right module structure to define a (right) C-algebra, and then extended the multiplication to make A a C - C bimodule, then the rules of our (left-handed) definition could have been deduced. In view of these remarks, we shall feel free to think of a C-algebra as a ringlet A which has a left or a right or a bimodule structure over C, satisfying the relevant subset of the rules shown above. That this is possible is a consequence of the commutativity of C, and the approach cannot in general be extended to cases in which commutativity is not available. Notice that in Chapter 2 we required that an algebra was a ring. It is traditional in representation theory not to make this restriction and we have followed that practice here.

When A is a ring with unity element $1 = 1_A$, then $1(cx) = (c1)x = c(1x) = cx = xc = (x1)c = x(1c) = (xc)1$ for any elements c of C and x of A. It is then routine to check that the map $\theta : C \to 1.c$ is a ring homomorphism of C into the centre of the ring A. Since a ring A is a bimodule over any of its subrings, this provides an alternative viewpoint on the bimodule structure of a

C-algebra A in the case that A is a ring. When a C-algebra is also a ring we shall say it is a unital C-algebra. If A and B are C-algebras then a C-algebra homomorphism is a ringlet homomorphism $f : A \to B$ such that $(cx)f = c(xf)$ for every c in C and x in A. The interested reader should check that, in the case in which A and B are rings, this definition implies that f is a ring homomorphism.

For an algebra A over an arbitrary commutative ring C, it need not be the case that A is (left) faithful over C, i.e. that $LA(A) = 0$. Even in the case that A is a ring, if the ring homomorphism of C into the centre of A has non-zero kernel N then $Nx = xN = 0$ for any x in A, and $N = LA(A) = RA(A)$. But if A is an algebra (not necessarily with 1) over a commutative field K then A must be (left) faithful, because $LA(A) = RA(A)$ is always an ideal $N \neq C$ of C, and the only ideal of K other than K itself is 0.

If C is a commutative ring and A is a unital C-algebra, let M be any right A-module. Then the product $mc = m(1_A c)$ is defined for any m in M and c in C, and it is trivial that M is a right C-module with this structure. The process of passing from the right A-module M_A to the right C-module M_C is called restriction of scalars, and could be carried out in the case of any ring-monomorphism $g : R \to A$ of a ring R into A. We shall use the same term to apply to that more general situation as and when the need arises. The analogous process involving left modules over A, C or R will be invoked without further comment.

The most important type of algebra for our present purposes is the group-algebra of a group G over a commutative ring C. In this construction, we take an arbitrary group C, and form the free left C-module $CG = \bigoplus_{g \in G} Cg$ on G. This is the module whose elements are formal sums $\sum_{g \in G} c_g g$, with $c_g = 0$ except for finitely many g in G, and with $\sum_{g \in G} c_g g = \sum_{g \in G} c'_g g$ if and only if $c_g = c'_g$ for each $g \in G$. Since C is commutative CG can be thought of as a C - C bimodule in the usual way, and we may then define a multiplication in CG according to the rule:

$$\left(\sum_{g\in G} c_g g\right)\left(\sum_{h\in G} c'_h h\right) = \sum_{f\in G}\left(\sum_{\substack{gh=f\\ g,h\in G}} c_g g c'_h h\right)$$

Since $cx = xc$ for any x in CG and any c in C, we have

$$\sum_{\substack{gh=f\\ g,h\in G}} c_g g c'_h h = \sum_{\substack{gh=f\\ g,h\in G}} c_g c'_h f = \left(\sum_{\substack{gh=f\\ g,h\in G}} c_g c'_h\right) f ,$$

so an equivalent definition of multiplication would be

$$\left(\sum_{g\in G} c_g g\right)\left(\sum_{h\in G} c'_h h\right) = \sum_{f\in G}\left(\sum_{\substack{gh=f\\ g,h\in G}} c_g c'_h\right) f .$$

It is an interesting but not difficult task to verify that, with C-module addition and this definition of multiplication, CG is an associative ring and a C-algebra. For example, if $k \in C$ and

$$x = \sum_{g\in G} c_g g, \quad y = \sum_{h\in G} c'_h h \in CG$$

then

$$k(xy) = k\left\{\sum_{f\in G}\left(\sum_{\substack{gh=f\\ g,h\in G}} c_g c'_h\right) f\right\}$$

$$= \sum_{f\in G}\left(\sum_{\substack{gh=f\\ g,h\in G}} (kc_g) c'_h h\right) f = \left(\sum_{g\in G} kc_g g\right)\left(\sum_{h\in G} c_h h\right)$$

$$= \sum_{f\in G}\left(\sum_{\substack{gh=f\\ g,h\in G}} c_g kc'_h\right) f = \left(\sum_{g\in G} c_g g\right)\left(\sum_{h\in G} kc_h h\right).$$

Hence $k(xy) = (kx)y = x(ky)$. Note that, to "move" k to the right of x we have used the fact that C is commutative.

We mentioned above that, by restriction of scalars, one can move from a module over a unital C-algebra to a module over C. In appropriate cases, it is possible to move in the opposite direction, from a module over a commutative ring C to a module over the group ring CG, for a group G. Thus, let M be a right C-module, say M_C, and let E be the ring of endomorphisms of M_C, with group of units Γ. Recall that we employ the convention that endomorphisms of a one-sided module are written on the opposite side to the scalars. Then the fact that M is a C - C bimodule can be described by the formula

$$(cm)d = c(md) \quad \text{for any } c, d \in C, \quad m \in M.$$

Hence left multiplication by any element of C is an endomorphism of the right C-module M, and it follows almost at once that the relevant map $C \to E$ is a ring homomorphism. Thus E too is a C - C bimodule, and we shall now show that $ce = ec$ for any $c \in C$, $e \in E$. For, if $m \in M$, we have

$$(ce)m = c(em) = (em)c = e(mc) = e(cm) = (ec)m,$$

so $ce = ec$. (Note that once again this depends upon the commutativity of C.) Hence the map of C into E is a map into the centre of E, so E is "naturally" a C-algebra. Furthermore, M is an E - C bimodule, from the definition of endomorphism, and the group of units Γ of E is the group of right C-module automorphisms of E.

Now suppose $f : G \to \Gamma$ is any group homomorphism of a group G to Γ. Let $f^* : CG \to E$ be the well-defined map which takes $\sum_{g \in G} c_g g \in CG$ to $\sum_{g \in G} c_g (gf) \in E$. This map is well defined, and since f is a homomorphism, C is a central subring of CG and $ce = ec$ for any $c \in C$, $e \in E$, it can easily be checked that f^* is a ring homomorphism. Since M is a left E module, it is a left $(CG)f^*$ module by restriction of scalars. Hence we can obtain a left CG-module structure on M by defining $xm = (xf^*)m$ for every $x \in CG$ and $m \in M$. Our previous concept of restriction of scalars involved thinking of a B-module as an A-module when A was (identified with) a subring of B. The construction we have just seen in the case of the ring homomorphism $f^* : CG \to E$, however, works for any ring homomorphism, and we will henceforth call the corresponding construction restriction of scalars. That is, if M is (say) a left B module and $h : D \to B$ is a ring homomorphism, we give M a left D-module structure by defining $dm = (dh)m$ for any $d \in D$ and $m \in M$. In the case that h is set-theoretic inclusion, this coincides with the previous concept of restriction of scalars.

Thus, given a group homomorphism $f : G \to \mathrm{Aut}(M_C)$ for some right module M over a commutative ring we obtain a ring homomorphism $f^* : CG \to \mathrm{End}(M_C)$ and a left CG-module structure on M. On the other hand, if $h : CG \to E = \mathrm{End}(M_C)$ is a ring homomorphism, then if $g \in G$

$$(1_C g)h \in \mathrm{Aut}(M_C) = U(\mathrm{End}(M_C))$$

(where $U(R)$ denotes the group of units of any ring R), because for any $g \in G$, $1_{CG} = 1_C 1_G = 1_C g 1_C g^{-1}$ so

$$(1_{CG})h = 1_E = (1_C g)h(1_C g^{-1})h .$$

Thus $(1_C g)h$ is a unit of E. For simplicity's sake we can identify C with the subset $C1_G$ of G, and G with the subset $\{1_C g : g \in G\}$ of $U(CG)$, and then what we have seen is that h induces, by restriction, a group homomorphism of G into $U(\mathrm{End}(M_C)) = \mathrm{Aut}(M_C)$.

We can thus define a representation of a group G over a commutative ring C to be a group homomorphism of G to the group of automorphisms of a C-module, or equivalently a ring homomorphism of the group algebra CG to the endomorphism ring of a C-module. Since C is commutative we could take either right or left C-modules; for the sake of definiteness we shall henceforth take right C-modules and continue writing the endomorphisms - and hence the action of CG - on the left. This approach will occasionally be modified, when we encounter a commutative subring L of CG. It would then be possible to think of M as a right L-module, but to do so would violate the convention that endomorphisms go on the opposite sides to scalars, and could create confusion.

Since the notions of a group homomorphism of a group G to $\mathrm{Aut}(M_C)$ and of a ring homomorphism of CG to $\mathrm{End}(M_C)$ are equivalent, it will cause no confusion to denote both the homomorphisms by the same symbol, say X. We shall then say that the module M <u>affords</u> the representation X. We shall chiefly be interested in the representations in which M is a finitely generated free C-module, that is $M \simeq C \oplus \ldots \oplus C$ (n copies) for some non-negative integer n. In this case it is well known (and easy to check) that $\mathrm{End}(M_C)$ is isomorphic to $M_n(C)$, the ring of $n \times n$ matrices over C. It would be possible (and is often done) to use the fact of this automorphism to define a representation as a ring homomorphism of CG into $M_n(C)$ for some n. The major disadvantage of so doing is that the isomorphism of $\mathrm{End}(M_C)$ and $M_n(C)$ is in general not unique. We shall, therefore, prefer to stick to the approach outlined so far, which is far more general in concept, and more in line with the mainstream of ring theory.

In the case that M is a finitely generated free C-module, $M \simeq C \oplus \ldots \oplus C$ (n copies), it can be shown that the integer n is unique. For if m is a maximal ideal of C, so that $C/m \simeq K$ is a field, then $n = \dim_K(M/Mm)$ where M/Mm is the factor C-module, which has a natural structure as a right K-vector space. Since, as is well known, the dimension of a (right) vector space is unique, it follows that n is unique. We shall then define the unique integer n to be the <u>degree</u> of the representation.

Suppose that $X : CG \to \mathrm{End}(M_C)$ and $Y : CG \to \mathrm{End}(N_C)$ are two representations of CG. They are said to be <u>similar</u> if there is a C-module isomorphism $T : M \to N$ (written on the left like an endomorphism) such that

$$T((xX)m) = (xY)(Tm)$$

for every x in CG and m in M. Note that any isomorphism $T : M \to N$ induces a ring isomorphism $T^* : \mathrm{End}(M) \to \mathrm{End}(N)$, defined by

$$(eT^*)n = (TeT^{-1})n$$

for every e in E and n in N. With this approach we can say that the two representations are similar if there is a C-module isomorphism $T : M \to N$ such that $XT^* = Y$.

REPRESENTATIONS OVER FIELDS

We shall be especially interested, in the applications which follow, in the case of representations of finite groups over fields. In such a case the group algebra KG is finite dimensional over the central subfield K. If N is any finitely generated left KG-module, then N is a homomorphic image of $KG \oplus \ldots \oplus KG$ (m copies) for some non-negative integer m, and each summand KG is finite dimensional so N is a finite dimensional left K-vector space by restriction of scalars, and hence a finite dimensional right K-vector space because K is commutative.

Let R be a ring. We recall from Chapter 2 that a one-sided R-module M is semisimple if it is the sum of its simple submodules, and that R is semisimple if R_R (equivalently ${}_R R$) is a semisimple R module. It was shown that R is semisimple if and only if it is a finite direct sum of full matrix algebras over division rings, or equivalently if R is a semiprime ringlet with either right or left d. c. c. A semisimple ring has a left and a right module composition series. One of the most important results of representation theory, known as Maschke's Theorem, asserts that in many cases the group algebra of a finite group over a field is a semisimple ring.

<u>4.1 Theorem.</u> Let K be a field and G a finite group, and suppose that the order $|G|$ of G is a unit of K. Then the group algebra KG is a semisimple ring.

<u>Proof.</u> Let $R = KG$. We need to show that ${}_R R$ is a semisimple R-module. From Chapter 2, it is sufficient to show that if V is a submodule of ${}_R R$ there is a submodule W such that ${}_R R = V \oplus W$. Since R is a left K-vector space, V is a K-subspace of R and there is a K-subspace N such that $V \oplus N = {}_K R$. Let $\pi : R \to R$ be the K-linear map which takes r to v, where $r = v + n \in R$ and $v \in V$, $n \in N$. Let $\sigma : R \to R$ be the map defined by

$$r\sigma = \sum_{g \in G} g^{-1}((gr)\pi) .$$

Certainly σ is a K-linear map of R to V. If $h \in G$ then

$$\begin{aligned}(hr)\sigma &= \sum_{g\in G} g^{-1}((ghr)\pi) \\ &= \sum_{g\in G} h(h^{-1}g^{-1})((ghr)\pi) \\ &= h \sum_{g\in G} (h^{-1}g^{-1})((ghr)\pi) \\ &= h(r\sigma) .\end{aligned}$$

It follows that σ is an R-homomorphism of R to V. Let $r\tau = \frac{1}{|G|} r\sigma$ for every $r \in R$. Then τ also is an R-homomorphism of R to V, and clearly $r\tau = r$ for every r in V, so $V = R\tau$. Let $\omega \in R$. Then

$$\omega\tau^2 = (\omega\tau)\tau = \omega\tau ,$$

so $\omega - \omega\tau \in \ker \tau$. Hence ${}_RR = V + \ker \tau$, and if $x \in V \cap \ker \tau$ then $x = x\tau = 0$, so the sum is direct as required. //

We shall need the following result.

<u>4.2 Lemma.</u> Let A be a ring and M a simple left A-module. Then M is isomorphic to ${}_A A/L$ for some left ideal L of A, and if A is semisimple then M is isomorphic to N for some left ideal N of A.

<u>Proof.</u> Let $0 \neq x \in M$. Clearly $M = Ax$, and the map $\theta : a \to ax$ is a surjective A-module homomorphism (an A-module <u>epimorphism</u>) with kernel $L = \{ y \in A : yx = 0 \}$, and it is then standard that θ induces an isomorphism ${}_A A/L \simeq M$. If A is semisimple there exists a left ideal N such that ${}_A A = N \oplus L$, and then the composite map $N \xrightarrow{\text{inclusion}} N \oplus L \xrightarrow{\theta} M$ is an isomorphism. //

If A is any semisimple ring then, as was shown in Chapter 2, every (left) A-module is a direct sum of simple A-modules. In representation theory, a simple module is often called <u>irreducible</u>, but that usage is in decline in ring theory. Since one of our aims is to provide a translation guide between the two areas, we have to tackle that problem. Since it is caused by different but widely prevalent usages among mathematicians, it appears better to highlight the discrepancy than to paper over it. We shall therefore continue to refer to <u>simple</u> modules, but shall say that a group representation which is afforded by a simple module is an <u>irreducible</u> representation, and (once we have met characters) we shall refer also to <u>irreducible</u> characters.

<u>4.3 Lemma.</u> Let A be a semisimple ring and V a left A-module. Let $E = \mathrm{End}_A(V)$ be the endomorphism ring of V, and suppose M is a simple submodule of V. Then if $H(M) = \Sigma W$, where W runs over all submodules isomorphic to M, then $H(M)$ is an $A - E$ subbimodule of V.

<u>Proof.</u> Let $f \in E$, and W be a submodule of V isomorphic to M. Then $Wf = 0$ or $Wf \simeq W$, so $W \subseteq H(M)$. Hence $H(M)E = H(M)$, as required. //

Let R be a ring and assume $R = A \oplus B$ where A and B are ideals of R. Then there exist $a \in A$ and $b \in B$ such that $1 = a + b$. Using the observation that $AB = 0$ it follows that a and b are central idempotents of R and that $A = Ra$ and $B = Rb$. If R is semisimple then it follows from the structure theorems proved in Chapter 2 that every ideal is generated by a central idempotent.

Using the chain conditions for semisimple rings it can be seen that an ideal is minimal if and only if it is generated by a central idempotent which cannot be written as the (direct) sum of two non-zero central idempotents. This observation enables us to see that any ideal is the direct sum of minimal ideals. These two comments are used in the next few results.

<u>4.4 Lemma.</u> Let A be a semisimple ring, M a simple A-module, and N a left ideal isomorphic to M. Then $H(N)$ is a minimal ideal of A.

<u>Proof.</u> If $f : A \rightarrow A$ is an endomorphism of the left A-module ${}_AA$ then, for any

$x \in A$, $xf = (x1)f = x(1f)$. Hence the endomorphisms of ${}_A A$ are simply the right multiplications by elements of A, so H(N) is an ideal. Since A is a direct sum of minimal ideals, so is H(N). Suppose H(N) contains two such minimal ideals; then it certainly contains a minimal ideal B such that $N \not\subseteq B$. Hence $B \cap N = 0$, so $BN = 0$. Hence $B(NA) = 0$. If L is a minimal left ideal contained in B then $LA = B$ since B is a minimal ideal, and $BL \neq 0$ since $B = B^2$. Hence $L \not\cong N$, contradicting the assumption that $B \subseteq H(N)$. Hence H(N) is a minimal ideal. //

If A is a semisimple ring and M is a simple submodule of the A-module V then the submodule H(M) identified in (4.3) is called the <u>homogeneous component</u> of M. It is implicit in the proof of (4.4) that there is one isomorphism class of simple A-modules for each minimal ideal of A, and there are no others. This observation immediately yields the following result.

<u>4.5 Lemma.</u> If A is a semisimple ring and V a non-zero left A-module there exist finitely many simple submodules of V, say $M_1, \ldots, M_r$, such that $V = H(M_1) \oplus \ldots \oplus H(M_r)$. //

If M is a simple module over a semisimple ring A, we denote by W(M) the homogeneous component H(N) where N is a simple left ideal of A isomorphic to M. The ideal W(M) is sometimes called the <u>Wedderburn component</u> of M in A. Note that, from the proof of Corollary (4.4), $W(M)N = 0$ for any simple A-module N such that $M \not\cong N$.

If A is any ring and V is any left A-module then V is naturally an A - E bimodule, where E is the endomorphism ring of V (over A). Since V is a right E-module, we can look at the endomorphism ring R of V over E. If $a \in A$ we let $\hat{a}$ denote the map from V to V defined by $\hat{a}(v) = av$ for every v in V. Since $(av)e = a(ve)$ for any a in A, v in V and e in E, $\hat{a}$ is an element of R. A routine verification shows that the map $a \to \hat{a}$ is a ring homomorphism of A to R, and trivial examples show that this map is not necessarily injective or surjective. In the case that A is a semisimple ring and M is a simple A module, however, it is possible to work out more exactly what is happening. The situation is described by the following result, which is known as the Double Centraliser Theorem.

4.6 Theorem. Let M be a simple module over a semisimple ring A, and let $D = \mathrm{End}_A(M)$ be the ring of A-endomorphisms of D. Let e be the central idempotent of A such that $W(M) = Ae$, and let $R = \mathrm{End}_D(M)$ be the ring of D-endomorphisms of M. Then the map $f : A \to R$, $a \to \hat{a}$ is surjective with kernel $A(1 - e)$.

Proof. Since $W(M) \cap (A(1 - e)) = 0$ it follows that $A(1 - e) \subseteq \ker f$; since $A = W(M) \oplus A(1 - e)$, so that $A(1 - e)$ is a maximal ideal of A, we must have $A(1 - e) = \ker f$. To show that f is surjective we first consider the special case of a simple left ideal N of A. We recall that $W(N) = NA$ is a minimal ideal of A, generated by the central idempotent e.

Since A is a semiprime ring, $0 \neq N^2 = N$. Thus we have a well-defined multiplication on N, which is a ringlet in its own right. If $m \in N$, we can define a map $d_m : N \to N$ by $xd_m = xm$ for any x in N. Let $a \in N$. Then $(ax)d_m = (ax)m = a(xm) = a(xd_m)$, so d_m is an A-endomorphism of N. Now let $w : N \to N$ be a D-endomorphism. Then for any m_1, m_2 in N we have

$$w(m_1 m_2) = w(m_1 d_{m_2}) = (wm_1) d_{m_2} = (wm_1) m_2 .$$

Now if $0 \neq n \in N$ then $W(N) = AnA$ so $e = \sum_{i=1}^{r} a_i n b_i$ for some a_i, b_i in A. Hence, if $m \in N$, $m = (\Sigma\, a_i n b_i) m$, so

$$\begin{aligned} wm &= w\{\Sigma (a_i n)(b_i m)\} \\ &= \Sigma (w(a_i n))(b_i m) \\ &= (\Sigma (w(a_i n)) b_i) m \\ &= \hat{v} m \end{aligned}$$

where $v = \Sigma (w(a_i n)) b_i \in A$. For the case of an arbitrary simple module M, let $u : M \to N$ be an A-module isomorphism of M with a left ideal N of A. If t is an $\mathrm{End}_A(M)$ endomorphism of M, then $utu^{-1} : N \to N$ is an $\mathrm{End}_A(N)$

endomorphism of N, and if $utu^{-1} = \hat{v} : N \to N$ then $t = \hat{v} : M \to M$ as required. //

If M is a (left) module over a ring R, with endomorphism ring E, then the endomorphism ring B(M) of the (right) module M_E is called the <u>bi-commutator</u> or <u>double centraliser</u> of M (over R). With this terminology, we have the following result.

<u>4.7 Corollary.</u> Let A be a semisimple ring and M a simple (left) A-module. Suppose the endomorphism ring of ${}_AM$ is D, a division ring by (2.14), then the module M, the division ring D and the bicommutator ring B(M) have the following properties:

(a) $B(M) \simeq W(M)$, the Wedderburn component of M in R;

(b) $\dim_D(M_D) = n < \infty$;

(c) $B(M) \simeq M_n(D)$, the full matrix algebra of $n \times n$ matrices over D.

<u>Proof.</u> That $B(M) \simeq W(M)$ is simply an immediate consequence of (4.6). Clearly (b) implies (c), so it will suffice to prove (b); in particular, it is sufficient to show that $\dim_D(M_D) < \infty$. Suppose otherwise; then $\{x \in B(M) : \dim_D(\mathrm{Im}\, x) < \infty\}$ is a non-zero ideal other than B(M). But B(M) is a simple ring by (2.23); contradiction. //

<u>CHARACTERS</u>

Let C be a commutative ring, n a natural number, and A, B be $n \times n$ matrices over C. The <u>trace</u> of A is the sum of the elements on its leading diagonal: formally, if $A = (a_{ij})$ where $a_{ij} \in C$, $1 \le i, j \le n$ then $\mathrm{Tr}(A) = \sum_1^n a_{ii}$. Since C is commutative, it is easily verified that $\mathrm{Tr}(AB) = \mathrm{Tr}(BA)$. Suppose B is invertible. Then $\mathrm{Tr}(B^{-1}AB) = \mathrm{Tr}((B^{-1}A)B) = \mathrm{Tr}(B(B^{-1}A)) = \mathrm{Tr}(IA) = \mathrm{Tr}(A)$. From this observation we can define the trace of certain representations.

For suppose $M \simeq C \oplus \dots \oplus C$ (n copies), that is M is free on n generators, and $f \in \mathrm{End}(M_C)$. The integer n is uniquely determined by M, as noted earlier in this Chapter. Any set of elements $x_1, \dots, x_m$ of M such that $M = x_1C \oplus \dots \oplus x_mC$ and each $x_iC \simeq C$ (as C module) is called a <u>free generating set</u> for M, and we must have $m = n$ from the uniqueness property just mentioned. Any C-module automorphism of M induces a permutation of the set of free generating sets, and any endomorphism f of M is an automorphism if and only if it is a permutation. Equivalently, f is an automorphism if and only if $\det(\tilde{f}) \in U(C)$, the group of units of C, for some matrix $\tilde{f}$ of f with respect to some free generating set of M. Given any two free generating sets of M, there is at least one automorphism which takes the one set to the other (so the group of automorphisms of M acts transitively on the set of free generating sets).

Thus, for any $f \in \mathrm{End}(M_C)$, we may define $\mathrm{Tr}(f) = \mathrm{Tr}(\tilde{f})$ where $\tilde{f}$ is the matrix of f with respect to some free generating set. Our previous remarks guarantee that Tr(f) is independent of the choice of free generating set.

Hence if $X : CG \to \mathrm{End}(M_C)$ is a representation (where M is finitely generated free over C), we have a well-defined element $\mathrm{Tr}(uX) \in C$ for any $u \in CG$. Such a function $X : CG \to C$ which takes $u \in CG$ to $\mathrm{Tr}(uX) \in C$ is a character on G (or on CG), said to be <u>afforded by</u> the representation X (or sometimes simply by the module M). The <u>degree</u> of a character is the integer n such that M is free on n generators, and a character is <u>irreducible</u> when the representation which affords it is. Two characters are <u>similar</u> if they are afforded by similar representations.

Let $X_1, \ldots, X_r$ be representations afforded by the C-modules $M_1, \ldots, M_r$. Then we have a map

$$X_1 \oplus \ldots \oplus X_r : CG \to \mathrm{End}(M_1 \oplus \ldots \oplus M_r)$$

where

$$u(X_1 \oplus \ldots \oplus X_r)(m_1 + \ldots + m_r) = (uX_1 m_1) + \ldots + (uX_r) m_r$$

for any u in CG and m_i in M_i, $1 \le i \le r$. It is easy to check that this map is also a representation, the <u>sum</u> of $X_1, \ldots X_r$. If each M_i is a finitely generated free C-module, and each X_i affords the character χ_i, then $X_1 \oplus \ldots \oplus X_r$ affords the character $\chi_1 + \ldots + \chi_r)$ since

$$\mathrm{Tr}(u(X_1 \oplus \ldots \oplus X_r)) = \mathrm{Tr}(uX_1) + \ldots + \mathrm{Tr}(uX_r) .$$

Hence we have a natural notion of addition of characters. Moreover, if m is a natural number and χ is a character, we can also define a multiplication $m\chi$ in the obvious way: $m\chi = \chi + \ldots + \chi$ (m summands).

We shall especially be interested in characters for finite groups over fields, but before going on to study them we need to consider group and ring homomorphisms.

CHANGE OF GROUP OR RING

Let B, C be commutative rings, and G, H be groups. If $f : B \to C$ is a ring homomorphism, it is trivial to verify that there is a unique ring homomorphism $f^* : BG \to CG$ such that $(\Sigma\, b_g g) f^* = \Sigma (b_g) fg$ for every element $\Sigma b_g g$ of BG. If $d : G \to H$ is a group homomorphism, it is equally trivial to verify that there is a unique B-algebra homomorphism $d^* : BG \to BH$ such that $(\Sigma b_g g) d^* = \Sigma b_g (g d^*)$ for every element $\Sigma b_g g$ of BG. We shall be concerned to look at the relations between BG- and CG-modules, or between BG- and BH-modules, in various situations which are special cases of one or other of those just described.

Our first result along these lines concerns normal subgroups, and involves a celebrated theorem due to Clifford [5].

4.8 Theorem. Let K be a field, G a group, F a normal subgroup of G, and M a simple (left) KG-module.

(a) Suppose M contains a simple KF-module. Then M is a semisimple KF-module, that is (2.24) M is a direct sum of simple KF-submodules.

(b) Suppose M satisfies d.c.c. on (left) KF-submodules. Then there are finitely many isomorphism classes of simple KF-submodules, with (say) $N_1, \ldots, N_s$ being a complete set of pairwise non-isomorphic representatives thereof. Furthermore, if for each i, $1 \leq i \leq s$, $M_i = \sum_{\substack{N \subseteq M \\ N \simeq N_i}}$ (where the sum runs over the simple KF-submodules of M) then each M_i is a finite direct sum of copies of N_i, and every simple submodule contained in M_i is isomorphic to

N_i. Moreover, G acts as a transitive permutation group on $M_1, \ldots, M_s$.

Proof. (a) Let W be a simple KF-submodule of M. Then $(KG)W = M$, so $M = \sum_{g \in G} gW$. If $h \in F$, $w \in W$, $g \in G$ then $hgw = gg^{-1}hgw = gh'w \in gW$ (for some $h' \in F$), so each gW is a KF-submodule of M, easily seen to be simple. Hence M is a sum of simple KF-modules as required.

(b) If M has d. c. c. on submodules then it contains no infinite direct sum of submodules (compare the proof of Theorem 2.23). Hence M is a finite direct sum of simple KF-modules, so M certainly has a composition series over KF. By Proposition 2.12, so does any KF-submodule L of M. By Proposition 2.13, the distinct isomorphism classes occurring in a composition series for L, and their respective frequencies, are uniquely determined by L. Conversely, if N is any simple submodule of any submodule L of M, it is trivial that N is a factor in a composition series for L. Statement (b) follows at once from these observations, except perhaps for the transitivity of G on $\{M_1, \ldots, M_s\}$. This latter follows, however, from the fact that $M = \sum_{g \in G} gL$ where L is any simple KF-submodule. //

There are examples of modules which satisfy condition (a) but not condition (b) of Clifford's theorem. For example, let $K \subseteq P$ be an infinite dimensional field extension, let $G = U(P)$ and $F = U(K)$. Then P is naturally a simple KG module, and is an infinite direct sum of simple KF modules. Most commonly, however, Clifford's theorem is proved under the assumption that $\dim_K M < \infty$, when (b) is automatically satisfied.

Let K be a field and A a finite-dimensional K algebra, with unity. Then K is a splitting field for A if, for each simple left A-module M, $\mathrm{End}_A(M) \simeq K$. (The inclusion $K \subseteq \mathrm{End}_A(M)$ follows from the commutativity of K.) If $A = KG$ is a group algebra for some finite group G, we say K is a splitting field for G.

4.9 Proposition. Let K be an algebraically closed field and A a finite-dimensional unital K-algebra. Then K is a splitting field for A.

Proof. Let M be a simple left A-module. Since $\dim_K A < \infty$ we have $\dim_K M < \infty$. Hence $\dim_K(\mathrm{End}_K(M)) < \infty$. But $D = \mathrm{End}_A(M) \subseteq \mathrm{End}_K(M)$, so $\dim_K \mathrm{End}_A(M) < \infty$. Now D is a division ring (2.14); suppose $d \in D$. Then $K \subseteq K[d] \subseteq D$, and $\dim_K K[d] \le \dim_K D < \infty$. Hence $K[d]$ is an integral domain (because D is), is commutative (because K is central in D), and satisfies d.c.c. on left ideals (because it is finite dimensional over K). Thus $K[d]$ is algebraic over K, and we have $K[d] \subseteq K$ because K is algebraically closed. Hence $D = K$. //

The importance of splitting fields is that they enable us to obtain additional information about semisimple algebras.

4.10 Theorem. Let K be a splitting field for a semisimple finite-dimensional K-algebra A. Let $N_1, \ldots, N_s$ be a complete set of pairwise non-isomorphic representatives of the isomorphism classes of simple left A-modules. Then

(a) for each i, $W(N_i) \simeq \mathrm{End}_K(N_i) \simeq M_{n_i}(K)$, where $n_i = \dim_K N_i$;

(b) $\dim_K W(N_i) = n_i^2$, $1 \le i \le s$;

(c) $W(N_i)$ has an A-composition series of length n_i;

(d) $\dim_K A = \sum_1^s n_i^2$;

(e) $\dim_K(Z(A)) = s$.

Proof. Theorem 2.23 gives us the basic structure of A as a direct sum of simple rings with d.c.c. on left and on right ideals. The assertions of the theorem then follow from Corollary 4.7 and the definition of a splitting field. //

We shall see later that, for suitable fields K and groups G, there is a finite-dimensional field extension $K \subseteq P$ such that the group algebra PG is semisimple and P is a splitting field for PG. First, however, we need to look at changes of group.

Let G be a group, H a subgroup, C a commutative ring, M a left CH-module. The induced module $M^G = CG \otimes_{CH} M$ (see Theorem 2.15) is a left CG-module, whose properties are related to those of M. Furthermore, if

$X : CH \to End(M_C)$ is a representation of H, the above construction yields a representation $XG : CG \to End(M^G)$, called the induced representation. In the case that M is a finitely generated free C-module, with one free generating set being $m_1, \ldots, m_n$, and that $[G : H]$ is finite, it is not difficult to check that M^G is a finitely generated free C-module; the interested reader may verify that, if $g_1, \ldots, g_r$ is a complete set of left coset representatives of H in G then $\{g_i \otimes m_j : 1 \le i \le r, 1 \le j \le n\}$ is a free generating set for M^G over C. Hence we have the induced character, which is afforded by the induced representation. If C is a field and G, H are finite groups then every C-module (i.e., C-vector space) is free, and if $\dim_C M < \infty$ then $\dim_C M^G = (\dim_C M)[G:H]$.

CHARACTER THEORY AT CHARACTERISTIC 0

We are now in a position to examine character theory in some more depth.

4.11 Proposition. Let C be a commutative ring, G a group and $X : CG \to End(M_c)$ be a representation of G which affords a character, say χ. If $g, h \in G$ and g, h are conjugate in G then $\chi(g) = \chi(h)$.

Proof. This follows at once from the fact that, if A, B are $n \times n$ matrices over C and if B is invertible then $Tr(A) = Tr(B^{-1}AB)$. //

We now fix the following notations for the results in the remainder of this section. Let G be a finite group, K a field, and suppose char K is 0, so that KG is semisimple, and every natural number is a unit of KG. Let $M_1, \ldots, M_s$ be a complete set of pairwise non-isomorphic representatives of the isomorphism classes of simple left KG-modules. Let $X_1, \ldots, X_s$ be the corresponding (irreducible) representations of G, and $\chi_1, \ldots, \chi_s$ the corresponding (irreducible) characters.

4.12 Proposition. With the above conventions, suppose also that K is a splitting field for KG. Then

(a) $|G| = \sum_1^s \chi_i(1)^2$;

(b) s is the number of distinct conjugacy classes of G.

Proof. It is clear that $\chi_i(1) = \dim_K M_i$, so (a) follows from Theorem 4.10. The same result states that $s = \dim_K Z(KG)$, so we need to show that $\dim_K Z(KG)$ is the number of distinct conjugacy classes of G. Suppose $H_1, \ldots, H_r$ are the distinct conjugacy classes, and for each j, $1 \le j \le r$, let $x_j = \sum_{h \in H_j} h$. If $g \in G$ then $gx_j = gx_j g^{-1} g = (\sum_{h \in H_j} ghg^{-1}) g = x_j g$, so $x_j \in Z(KG)$. Clearly $x_1, \ldots, x_r$ are linearly independent so $r \le \dim_K Z(KG)$. In the other direction, suppose $w = \sum_{h \in G} a_h h \in Z(KG)$. Then $gwg^{-1} = w$ for each $g \in G$, so if $h_1, h_2 \in H_j$ for some j we must have $a_{h_1} = a_{h_2}$. Hence $r \ge \dim_K Z(KG)$, so finally $r = \dim_K Z(KG) = s$. //

For example, if $G = S_3$, the permutation group on three points, then G has three conjugacy classes, of orders 1, 2 and 3. If K is the rational field then K is a splitting field for KG, and $KG \simeq K \oplus K \oplus M_2(K)$, (Exercise!).

If F is an algebraically closed field of characteristic 0, and G is any finite group, then F is a splitting field for FG (4.9), so G is abelian if and only if all its irreducible representations over F have degree 1. This last statement is not necessarily true if F is not assumed to be algebraically closed. For example, let C be a cyclic (multiplicative) group of order 3, generated by x, and let K be the rational field, and $w = e^{2\pi i/3}$ be a complex cube root of 1. Then KC is commutative and $e = \frac{1}{3}(1 + x + x^2)$ is a non-trivial central idempotent, and KC/KCe is isomorphic to the field $K[w]$. Hence $KC \simeq K \oplus K[w]$, so C has an irreducible representation of degree $\dim_K K[w] = 2$. Note that C is a subgroup of S_3 so that a splitting field for a finite group G is not necessarily a splitting field for a subgroup.

We now return to the standard conventions of this section. The (left) regular representation of G is the representation in which elements of g act on KG_K by left multiplication, and the character afforded by this representation is the (left) regular character of G.

4.13 Lemma. Let ρ be the regular character of G then, for g in G, $\rho(g) = |G|\delta_{g1}$, where $\delta_{g1} = 0$ or 1 according as $g \neq 1$, $g = 1$.

Proof. Choose the elements of G as a K-basis of KG, say $g_1, \ldots, g_n$. If $g \in G$, $g \neq 1$ then $gg_i = g_j$ for some $j \neq i$, so the matrix of (left multiplication by) g with respect to the basis $g_1, \ldots, g_n$ has as its only non-zero entries 1s off the leading diagonal. If $g = 1$, clearly $\rho(g) = |G| = \dim_K KG$. //

Recall that a sum of characters is also a character, as is the product of a character by a positive integer.

4.14 Lemma. Let ρ be the regular character and K, G, $\chi_1, \ldots, \chi_s$ have their usual meanings. If K is a splitting field for KG then

$$\rho = \sum_1^s \chi_i(1)\chi_i .$$

Proof. If $A = KG$ then ${}_A A = \oplus W(M_i)$, and (4.10) each $W(M_i)$ is the direct sum of $\dim_K M_i = \chi_i(1)$ copies of M_i. //

4.15 Lemma. Let K, G, $\chi_1, \ldots, \chi_s$ have their usual meanings and let K be a splitting field for KG. Suppose that, for $1 \leq i \leq s$, e_i is the central idempotent of KG such that $W(M_i) = KGe_i$. Then

$$e_i = \frac{1}{|G|} \sum_{h \in G} \chi_i(1)\chi_i(h^{-1})h .$$

Proof. Suppose $e_i = \sum_{g \in G} a_g g$ where each $a_g \in K$. Since $\rho(x + y) = \rho(x) + \rho(y)$ for any elements x, y of KG, we have $\rho(e_i h^{-1}) = a_h|G|$ by Lemma 4.13. Hence

$$e_i = \sum_{h \in G} a_h h = \frac{1}{|G|} \sum_{h \in G} \rho(e_i h^{-1})h$$

$$= \frac{1}{|G|} \sum_{h \in G} \sum_{j=1}^{s} \chi_j(1)\chi_j(e_i h^{-1})h .$$

Since e_i acts as the identity on $W(M_i) = KGe_i$, and annihilates $(1 - e_i)KG$, we have

$$e_i = \frac{1}{|G|} \sum_{h \in G} \chi_i(1)\chi_i(h^{-1})h. \ //$$

The above result leads us to the important Orthogonality Relations, which are stated in the next two results.

<u>4.16 Theorem.</u> Let K, G, $\chi_1, \ldots, \chi_s$ have their usual meanings, and K be a splitting field for G. If $h \in G$ then

$$\frac{1}{|G|} \sum_{g \in G} \chi_i(gh)\chi_j(g^{-1}) = \delta_{ij} \frac{\chi_i(h)}{\chi_i(1)} .$$

<u>Proof.</u> In the notation of Lemma 4.15, we know $e_i e_j = \delta_{ij} e_i$ (where δ_{ij} is the usual Kronecker delta), so

$$\left(\frac{1}{|G|} \sum_{g \in G} \chi_i(1)\chi_i(g^{-1})g\right) \left(\frac{1}{|G|} \sum_{g \in G} \chi_j(1)\chi_j(g^{-1})g\right)$$

$$= \frac{\delta_{ij}}{|G|} \sum_{h \in G} \chi_i(1)\chi_i(h^{-1})h.$$

Comparing coefficients of h,

$$\frac{\delta_{ij}}{|G|} \chi_i(1)\chi_i(h^{-1}) = \frac{1}{|G|^2} \sum_{k \in G} \chi_i(1)\chi_i(kh^{-1})\chi_j(1)\chi_j(k^{-1})$$

so

$$\frac{1}{|G|} \sum_{k \in G} \chi_i(kh^{-1})\chi_j(k^{-1}) = \frac{\delta_{ij}\chi_i(h^{-1})}{\chi_i(1)}$$

which is equivalent to the desired result. //

<u>4.17 Corollary.</u> If K, G, $\chi_1, \ldots, \chi_s$ have their meanings in (4.16) then

$$\frac{1}{|G|} \sum_{g \in G} \chi_i(g)\chi_j(g^{-1}) = \delta_{ij} .$$

Proof. Put $h = 1$ in (4.16). //

This last result is a classical orthogonality relation, and Theorem 4.16 is the General Orthogonality Relation.

COMPLEX REPRESENTATIONS

In this section we assume that K is a subfield of the complex field, that G is a finite group and that K is a splitting field for every subgroup H of G. (We have already seen, p. 93, that this is a strictly stronger condition than simply assuming K is a splitting field for G.) We let $M_1, \ldots, M_s$ be a complete set of pairwise non-isomorphic simple left KG-modules, $X_1, \ldots, X_s$ be the corresponding representations, and $\chi_1, \ldots, \chi_s$ be the corresponding characters.

4.18 Lemma. Let M be a left KG-module which affords the representation X and the character χ, and let $g \in G$. Suppose $H = \langle g \rangle$, the subgroup generated by g, and let $\chi(1) = f = \dim_K M$ and $|H| = n$. Then

(a) There is a K-basis of M with respect to which gX has a diagonal matrix, with entries say $Z_1, \ldots, Z_f$;

(b) Each Z_i is an nth root of 1;

(c) $\chi(g) = \sum_1^f Z_j$;

(d) $\chi(g^{-1}) = \overline{\chi(g)}$, the complex conjugate of $\chi(g)$.

Proof. Since K is of characteristic 0, KH is a commutative semisimple ring, i.e. a finite direct sum of fields. Since K is a splitting field for H, each of these fields is isomorphic to K, by Theorem 4.10. By Maschke's theorem (4.1), M is a finite sum of simple KH-modules. Since each such simple module has just been shown to have dimension 1 over K, there are f distinct simple summands, and if $x_1, \ldots, x_f$ are their KH-generators then $M = Kx_1 \oplus \ldots \oplus Kx_f = x_1K \oplus \ldots \oplus x_fK$. Since g is an automorphism of M, and $g^n = 1$, (a), (b) and (c) follow at once.

For (d), note that if a is a complex number and $|a| = 1$ then

$$|a|^2 = a\bar{a} = 1^2 = 1,$$

so $a^{-1} = \bar{a}$. Since g^{-1} clearly has a diagonal matrix with respect to $x_1, \ldots, x_f$ whose entries are $Z_1^{-1}, \ldots, Z_f^{-1}$, we have

$$\chi(g^{-1}) = \sum_1^f Z_i^{-1} = \sum_1^f \bar{Z}_i = \overline{(\sum_1^f Z_i)} = \overline{\chi(g)}. \ //$$

4.19 Corollary. With the situation of Lemma 4.18, $\chi(g)$ is an algebraic integer for any g in G.

Proof. Each root of 1 is an algebraic integer, so $\chi(g)$ is a sum of algebraic integers. From Lemma 2.8, $\chi(g)$ is an algebraic integer. //

The set Map(KG, K) of functions from KG to K is naturally a (left) K-vector space, so we may consider the subspace Ch(KG, K) which is generated over K by the irreducible characters. Since KG is semisimple, every character χ is a sum $\chi = \sum_{i=1}^{s} a_i \chi_1$ for some non-negative integers a_i (because every module is semisimple), and hence an element of Ch(KG, K). It is natural to ask, what is $\dim_K$Ch(KG, K) ?

4.20 Proposition. With the notation just explained, $\dim_K$Ch(KG, K) = s, or equivalently $\chi_1, \ldots, \chi_s$ are linearly independent over K.

Proof. Suppose $a_1, \ldots, a_s \in K$ and $\sum_1^s a_i \chi_i = 0$. Pick j, $1 \leq j \leq s$, and consider the Wedderburn component $W_j = W(M_j)$ of KG. Since K is a splitting field for G, $\dim_K M_j = n_j < \infty$ and Theorem 4.10 implies that $W_j \simeq M_{n_j}(K)$, the ring of $n_j \times n_j$ matrices over K. Let $x \in W_j$ be such that, with respect to some K basis of M_j, x has a matrix with only one non-zero entry, on the leading diagonal. Then

$$0 = \sum_1^s a_i \chi_i(x) = a_j \chi_j(x),$$

so $a_j = 0$. //

Suppose that, in addition to our standard hypotheses, K is invariant under complex conjugation: $K = \{\bar{x} : x \in K\}$. Then Corollary 4.17 may be reinterpreted in terms of the vector space $Ch = Ch(KG, K)$. If $v = \sum_1^s v_i \chi_i$ and $w = \sum_1^s w_i \chi_j$ are elements of Ch, for some $v_i, w_j \in K$, we can define a scalar product

$$(v, w) = \sum_{i,j=1}^{s} v_i \bar{w}_j \left(\frac{1}{|G|} \sum_{g \in G} \chi_i(g) \chi_j(g^{-1})\right)$$

$$= \sum_{i,j=1}^{s} v_i \bar{w}_j \left(\frac{1}{|G|} \sum_{g \in G} \chi_i(g) \overline{\chi_j(g)}\right).$$

Since $K = \{\bar{x} : x \in K\}$ it is easy to check that $(v, w) \in K$ for any $v, w \in Ch$. It is also straightforward to check (using Proposition 4.20) that this product is a hermitian, conjugate bilinear positive definite inner product (in the sense of the definitions and axioms given for complex vector spaces in [10 , Chapter 3]). It follows that

$$(\chi, \psi) = \sum_{g \in G} \chi(g) \psi(g^{-1}) = \sum_{g \in G} \chi(g) \overline{\psi(g)}$$

for any two characters χ, ψ on KG.

If K is not invariant under complex conjugation, i.e. if $K \neq \{\bar{x} : x \in K\}$, then (v, w) need not (from our definition) be an element of K. It must, however, be an element of the complex field $\mathbb{C}$, and in fact of the field $K(\bar{K})$ generated over K by $\bar{K} = \{\bar{x} : x \in K\}$. This makes the axiomatic characterisation of (,) a bit more messy, but does not affect anything essential. We shall use (,) as a complex-valued product on Ch(KG, K).

<u>4.21 Theorem.</u> Let G be a finite group, and K a subfield of the complex field. Suppose that K is a splitting field for G, and that M is a simple KG-module which affords the representation X and the character χ. Then $\chi(1) \mid |G|$. (Note K is not taken to be a splitting field for $H \leq G$.)

Proof. Let $H_1, \ldots, H_s$ be the distinct conjugacy classes of G. By Proposition 4.12 $s = \dim_K Z(KG)$, and from the proof of that result $Z(KG)$ has a K-basis $x_1, \ldots, x_s$, where $x_j = \sum_{h \in H_j} h$. If $z \in Z(KG)$ then, in the notation of the Double Centraliser Theorem, $zX = \hat{w}$ for some unique root of 1 $w \in K$, or $zX = \hat{0}$, according as $ze_j \neq 0$ or $= 0$, where e_j is the central idempotent such that $W(M) = KGe_j$. If $g, h \in H_j$ then $\chi(g) = \chi(h)$ (4.11), so $\chi(g)|H_j| = \chi(x_j) = \chi(1)w_j$, where $x_jX = \hat{w}_j$ for $w_j \in K$. Since $(\chi, \chi) = 1$, we have $\sum_{g \in G} \chi(g)\chi(g^{-1}) = |G|$. For each j, let $g_j \in H_j$. Then

$$|G| = \sum_{j=1}^{s} |H_j|\chi(g_j)\chi(g_j^{-1})$$

$$= \sum_{j=1}^{s} \chi(1)w_j\chi(g_j^{-1})$$

so

$$|G|\{\chi(1)\}^{-1} = \sum_{j=1}^{s} w_j\chi(g_j^{-1}) .$$

Since the right hand side is an algebraic integer and the left hand side is a rational number, both must be rational integers. That is, $\chi(1) \mid |G|$. //

INDUCED CHARACTERS

Let K be a field, G a group, H a subgroup, M a KH-module affording the representation X and the character χ. We have already met the induced module

$$M^G = KG \otimes_{KH} M$$

and the induced representation X^G and induced character χ^G which it affords. There are some pleasing relations between χ and χ^G. The first is given by the following result.

4.22 Proposition. In the above situation, suppose G is finite and $|G|$ is a unit of K. Then, if $g \in G$,

$$\chi^G(g) = \frac{1}{|H|} \sum_{x \in G} \dot{\chi}(xgx^{-1})$$

where $\dot{\chi}(f)$ is the function defined by $\dot{\chi}(f) = \chi(f)$ if $f \in H$, and $\dot{\chi}(f) = 0$ otherwise.

<u>Proof.</u> Note first that, since $|G|$ is a unit of K, so also is $|H|$, so the right hand expression makes sense. Since $\dim_K M < \infty$ we can find a basis $m_1, \ldots, m_n$ of M over K. Let $t_1, \ldots, t_r$ be a complete set of left coset representatives of H in G (that is, a left transversal to H in G). Then $\{t_i \otimes m_j : 1 \le i \le r, 1 \le j \le n\}$ is a K-basis of M^G, and for each pair i, j and each $g \in G$,

$$\begin{aligned} g(t_i \otimes m_j) &= t_k h \otimes m_j \\ &= t_k \otimes hm_j \end{aligned}$$

for some $h \in H$ and some $k, 1 \le k \le r$. If $\hat{g}$ is the matrix of g with respect to $\{t_i \otimes m_j\}$ then the above expression contributes 0 to $\chi^G(g) = \mathrm{Tr}(\hat{g})$ unless $k = i$. In that case we have

$$gt_i = t_i h.$$

If we fix i and consider the basis elements $t_i \otimes m_1, \ldots, t_i \otimes m_n$ then we obtain an $n \times n$ "block" in the matrix of $\hat{g}$, namely the matrix of h with respect to $m_1, \ldots, m_n$. Hence

$$\begin{aligned} \chi^G(g) &= \sum_{t_j} \dot{\chi}(t_j g t_j^{-1}) \\ &= \frac{1}{|H|} \sum_{x \in G} \dot{\chi}(xgx^{-1}), \end{aligned}$$

as required. //

Suppose that K is a subfield of the complex field, and that K is a splitting field for any subgroup H of the finite group G. Then we saw above that it is possible to define a complex-valued inner product $(\ ,\ )_F$ on the space Ch(KF, K) for any subgroup F of G. We saw also that the irreducible characters form a K-basis of Ch(KF, K), which contains all the characters of F over K. Clearly, if χ is a character of KG, and H is a subgroup of G, then by restriction we obtain a character of KH. We denote this restricted character by $\chi|_H$. The following important result of Frobenius connects the results of inducing and of restricting characters.

4.23 Theorem. Let K be a subfield of the complex field and G be a finite group. Suppose that K is a splitting field for each subgroup H of G. Let χ be a character of G and ψ be a character of H, and $(\ ,\ )_G$, $(\ ,\ )_H$ denote the standard inner products. Then

$$(\chi, \psi^G)_G = (\chi|_H, \psi)_H.$$

Proof. It is a straightforward consequence of the definitions and of Proposition 4.20 that

$$(\chi, \psi^G) = \frac{1}{|G|} \sum \chi(x)\psi^G(x^{-1}),$$

so from Proposition 4.22

$$\begin{aligned}
(\chi, \psi^G) &= \frac{1}{|G||H|} \sum_{x \in G} \chi(x) \Big(\sum_{y \in G} \dot{\psi}(yx^{-1}y^{-1}) \Big) \\
&= \frac{1}{|G||H|} \sum_{x, y \in G} \chi(yxy^{-1}) \dot{\psi}(yx^{-1}y^{-1}) \\
&= \frac{1}{|H|} \sum_{z \in G} \chi(z) \dot{\psi}(z^{-1}) \\
&= \frac{1}{|H|} \sum_{z \in H} \chi(z) \psi(z^{-1}) \\
&= (\chi|_H, \psi)_H. \quad //
\end{aligned}$$

CHANGE OF RING

Our ultimate objective in this section is to demonstrate relations between the representations of finite groups over certain algebraic number fields and their representations over certain finite fields. We begin by showing that, if G is a finite group, there is an algebraic number field K, finite dimensional over the rational field $\mathbb{Q}$, such that K is a splitting field for every subgroup H of G.

4.24 Theorem. Let G be a finite group and P a subfield of the complex field $\mathbb{C}$. Then there are fields $L, K \subseteq \mathbb{C}$ such that $P \subseteq L \subseteq K$ and $\dim_P K < \infty$ and (a) L is a splitting field for G, and (b) K is a splitting field for every subgroup H of G.

Proof. Since G has only finitely many subgroups, it clearly suffices to prove (a). Since (p. 54) a complex number is algebraic over P if and only if it is integral over P, it suffices to show that there is a finite set $S \subseteq \mathbb{C}$ such that each element of S is algebraic over P and the field $L = P(S)$ generated by S and P is a splitting field for G. For the argument in the proof of Corollary 2.6, plus induction, will then show that $\dim_P L < \infty$ as required.

Since $\mathbb{C}$ is algebraically closed, the integral closure A of P in $\mathbb{C}$ is a field (Lemma 2.9(b)) which is also algebraically closed, and hence a splitting field for G. Thus $AG \simeq \bigoplus_{i=1}^{s} M_{n_i}(A)$ for some $s \geq 1$ and some natural numbers $n_1, \ldots, n_s$. In $\bigoplus_{i=1}^{s} M_{n_i}(A)$ let ${}^{*}E^r_{hj}$ denote the element of $M_{n_r}(A)$ which has entries $\delta_{ph}\delta_{qj}$, $1 \leq p, q \leq n_r$, and let E^r_{hj} denote its inverse image in AG (under the selected isomorphism). Then $\{E^r_{hj}\}$ is an A-basis of G, and the elements multiply by the rule:

$$E^r_{hj}E^t_{km} = E^r_{hm}\delta_{rt}\delta_{jk}.$$

For each $g \in G$ there exist finitely many elements $a^r_{hj}(g) \in A$ such that $g = \sum_{h,j,r} a^r_{hj}(g) E^r_{hj}$. Conversely, for each E^r_{hj} there exist elements $b^r_{hj}(f) \in A$ such that

$$E^r_{hj} = \sum_{f \in G} b^r_{hj}(f) f .$$

If $S = \{a^r_{hj}, b^r_{hj}\}$ then S is a finite set. Let $L = P(S)$, then as already remarked $\dim_P L < \infty$, and it is easy to check that $LG = \sum_{h,j,r} LE^r_{hj}$. Since LG contains the elements $\{E^r_{hj}\}$, we have

$$LG \simeq \bigoplus_{L=1}^{s} M_{n_i}(L) ,$$

so that (compare the proof of Theorem 4.10) L is a splitting field for LG, and hence for G, as required. //

<u>4.25 Corollary.</u> Let G be a finite group. There is a finite-dimensional algebraic extension L of the rational field $\mathbb{Q}$ such that L is a splitting field for G. //

Let R be a commutative integral domain with unity element. Then R has a field of fractions K, by a familiar process which is a special case of the construction of rings of fractions as discussed in Chapter 2 (p. 25 ff). If G is a group and M is a KG-module then M is naturally an RG-module by restriction of scalars. Suppose that M is finite-dimensional over K, with basis say $m_1, \ldots, m_t$, and that G is finite. Let

$$W = \sum_{i=1}^{t} \sum_{g \in G} Rgm_i .$$

Then W is a finitely generated R-module (on the left and the right, because R is commutative). If W is also a free (right) R-module, then we have a representation of RG which will afford a character of G.

It should be pointed out that W is not necessarily a free R-module. Let $\mathbb{Z}$ be the ring of rational integers, $\mathbb{Q}$ the rational field, and $K = \mathbb{Q}[i]$ where $i^2 = -1$. If $R = \mathbb{Z}[2i]$ then R is a commutative integral domain with field of fractions K. Suppose G is a cyclic group of order 4, generated by g, and that M is the KG module given on K by defining $gw = iw$ for $w \in K$. If we choose i as a K-basis of M, it is a non-trivial (but not very difficult) task to show that the RG-module $W = \sum_{g \in G} Rgi$ is not free over R.

Nevertheless, for a suitable ring R and a suitable basis $m_1, \ldots, m_t$ of M over K, we can arrange that W is free over R, and even that W has a free generating set of t elements. The case of interest here is given by the following result.

4.26 Theorem. Let R be a commutative principal ideal domain with unity, and let K be its quotient field. Let G be a finite group and M a finitely generated (left) KG-module. Suppose $\dim_K M = t$. Then there exists an RG-submodule W of M, with a free generating set $x_1, \ldots, x_t$ of t elements over R, such that $M = \sum_1^t x_i K$.

Proof. Let $m_1, \ldots, m_t$ be a K-basis of M, and $W = \sum_{g \in G} \sum_{j=1}^{t} Rgm_i$, as above. Then (as before) W is an RG submodule of M finitely generated as R-module, and it is clear that

$$M = \sum_1^t m_i K = WK .$$

Since $ax = 0$, $a \in K$, $x \in M$ implies $a = 0$ or $x = 0$, W inherits a similar property: $ry = 0$, $r \in R$, $y \in W$ implies $r = 0$ or $y = 0$. That is, W is a finitely generated torsion-free R module, so by a fundamental result we have used several times before, W has a free generating set of n elements, say $x_1, \ldots, x_n$. Since $M = WK = \sum_1^n x_i K$, $n \geq t$. We claim $x_1, \ldots, x_n$ are linearly independent over K. For suppose $x_1 a_1 + \ldots + x_n a_n = 0$, where $a_1, \ldots, a_n \in K$. There exists $0 \neq b \in R$ such that each $a_i b \in R$ (by a common denominator argument), so $x_1(a_1 b) + \ldots x_n(a_n b) = 0$. Hence each $a_i b = 0$, and since $b \neq 0$ we must have $a_i = 0$ for each i. Hence $n \leq t$, and therefore $n = t$, as required. //

Now let L be an algebraic number field finite-dimensional over $\mathbb{Q}$, with ring of algebraic integers A. Let p be a rational prime and let B be a prime ideal of A such that $B \cap \mathbb{Z} = p\mathbb{Z}$ (2.11). The field $F = A/B$ has characteristic p, and is finitely generated over its prime field (p. 58) since A is finitely generated over $\mathbb{Z}$. Thus F is a finite field, say $F \simeq GF(p^f)$ for some $f \geq 1$. We met examples of this situation on pages 63 and 64. If $n \geq 1$ is a natural number, and R is a ring, I an ideal of R, let $M_n(I)$ denote the $n \times n$ matrices over A all of whose entries are in I. Then there is a ring epimorphism $\pi : M_n(R) \to M_n(R/I)$ with kernel $M_n(I)$, where the action of π is to take each matrix X in $M_n(R)$ to the element of $M_n(R/I)$ whose entries are those of X reduced mod I.

<u>4.27 Theorem.</u> Let G be a finite group and L be an algebraic number field finite-dimensional over $\mathbb{Q}$ which is a splitting field for G. Suppose A is the ring of algebraic integers of L, and p is a rational prime such that $p \nmid |G|$. Let B be a prime ideal of A such that $B \cap \mathbb{Z} = p\mathbb{Z}$. Then $F = A/B$ is a splitting field for G, and if

$$LG \simeq \bigoplus_{i=1}^{s} M_{n_i}(L)$$

then

$$FG \simeq \bigoplus_{i=1}^{s} M_{n_i}(F) .$$

<u>Proof.</u> Let R denote the local ring of A at B, and P its maximal ideal. The composite map $A \to R \to (R/P)$ induces an embedding $(A/B) \to (R/P)$ which is easily checked to be an isomorphism. Suppose $e_1, \ldots, e_s$ are the central idempotents of LG which generate the Wedderburn components of LG. From (4.15)

$$e_i = \frac{1}{|G|} \sum_{h \in G} \chi_i(1) \chi_i(h^{-1}) h$$

where, $1 \leq i \leq s$, χ_i is the character corresponding to e_i. Now $\chi_i(g) \in A$ for $g \in G$, and $|G|^{-1} \in R$ since $p \nmid |G|$. Hence, for each i, $e_i \in RG$, the subring of LG generated by R and G. It is clear that $RG/PRG \simeq (R/P)G = FG$; denote it by S. Let $\bar{e}_i$, $1 \leq i \leq s$, denote the image of e_i in S. Then (identifying S with FG) the coefficient of 1 in the expression for $\bar{e}_i$, $\bar{e}_i = \sum_{g \in G} a_g g$ for some $a_g \in F$, is the image of $\chi_i(1)^2 |G|^{-1}$, which is non-zero since $\chi_i(1) \,\big|\, |G|$ (4.21) and $p \nmid |G|$. Thus each of $\bar{e}_1, \ldots, \bar{e}_s$ is a non-zero element of FG. Now $e_1, \ldots, e_s$ are pairwise orthogonal ($e_i e_j = e_i \delta_{ij}$, $1 \leq i, j \leq s$), central idempotents whose sum is 1, so the same is true of $\bar{e}_1, \ldots, \bar{e}_s$ in FG. It follows from the proof of (4.12) that, for any field P and any finite group H, $\dim_P Z(PH)$ is the number of distinct conjugacy classes of H. From (4.12), $\dim_K Z(KG) = s$ is the number of distinct conjugacy classes of G. Hence $\bar{e}_1, \ldots, \bar{e}_s$ is an F-basis of $Z(FG)$. Pick j, $1 \leq j \leq s$, and let M be the simple KG-module whose Wedderburn component is $e_j KG$. From page 63 R is a principal ideal domain (with unity) and then by (4.26) M contains an RG submodule L free on $n = \dim_K M$ generators. Let $N = L/LP$. It is clear that N is an S module, and $\dim_F N = n$. Since $e_i M = 0$, if $1 \leq i \leq s$ and $i \neq j$, we have $\bar{e}_i N = 0$, for $1 \leq i \leq s$ and $i \neq j$. Hence N is a finite direct sum of isomorphic simple FG modules, say $N \simeq N_j \oplus \ldots \oplus N_j$ (r_j copies), where $\dim_F N_j = m_j = n/r_j$. Furthermore, since $N_1, \ldots, N_s$ are pairwise non-isomorphic simple FG modules, they represent all the isomorphism classes. For each j, let $D_j = \mathrm{End}_{FG}(N_j)$. By Wedderburn's Theorem (3.26), D_j is commutative. But by (4.7),

$$FG \simeq \bigoplus_{i=1}^{s} M_{m_j}(D_j), \text{ so } Z(FG) \simeq \bigoplus_{i=1}^{s} D_j.$$

Since $\dim_F Z(FG) = s$, we must have $D_j \simeq F$, $1 \leq j \leq s$. That is, F is a splitting field for G, and

$$FG \simeq \bigoplus_{i=1}^{s} M_{m_j}(K),$$

where $m_j \,|\, n_j = \chi_j(1)$, $1 \leq j \leq s$. Since $\sum_1^s n_i^2 = |G|$ by (4.17), we must have $m_j = n_j$, $1 \leq j \leq s$, as required. //

APPLICATIONS OF CLIFFORD'S THEOREM

In this section, we obtain a detailed description of what happens when a group G of automorphisms of a finite-dimensional vector space contains a (non-trivial) abelian normal subgroup.

4.28 Theorem. Let K be a commutative field and M a (right) K-vector space. Suppose that $E = \mathrm{End}_K(M)$ and A is an abelian subgroup of $U(E)$. Let KA be the group ring of A over K, and C be its homomorphic image generated in E by A and K (acting on the left). Suppose M is a simple KA module. Then

(a) C is a field, and $\dim_K C = \dim_K M$. Suppose also $|A| = t < \infty$. Then

(b) A is cyclic and $\dim_K M < \infty$;

(c) C is the splitting field for $x^t - 1$ over K;

(d) C is a Galois extension of K, and $\mathrm{Gal}(C/K)$ is a subgroup of $U(\mathbb{Z}/t\mathbb{Z})$, the multiplicative group mod t of integers prime to t; thus $|G| \mid \phi(t)$;

(e) if K is of characteristic $p \neq 0$ and K contains only finitely many roots of 1 then there exist integers $n, f \geq 1$ such that

(i) K contains exactly $p^f - 1$ distinct roots of 1, and

(ii) $n = \dim_K C$ is the least natural number such that $t \mid (p^{nf} - 1)$.

Proof. Since M is a simple KA-module it is a simple C-module, say $M = Cx$ where $0 \neq x \in M$. Now $D = \mathrm{End}_C(M)$ is a division ring, by Schur's Lemma (2.14). Let $d \in D$ and $W = cx \in M$ (where $c \in C$). Then $wd = (cx)d = c(xd) = c(ax)$ for some $a \in C$. It is easy to check that the map $d \to a$ is well-defined and is a ring isomorphism $D \simeq C$. Hence D is commutative and C is a commutative division ring, i.e. a field. Since evidently $C \simeq M$ as C-modules, $\dim_K C = \dim_K M$. This proves (a).

So suppose henceforth that A is finite, and $|A| = t$. Then $\dim_K KA = t < \infty$, and M is simple, so certainly is a finitely generated KA module. Hence $\dim_K M = n < \infty$. Since A is a finite group of units of the field C, A is cyclic (p. 58). This proves (b).

Now let B be the group of tth roots of 1 in C. Certainly $A \subseteq B$. If $A \neq B$, let $u \in B$, $u \notin A$. Then since B is abelian, the subgroup A^* generated by A and u is a homomorphic image of the (external) direct product $A \times \langle u \rangle$, so is finite, hence also cyclic. But every element of A^* satisfies $x^t = 1$, so $|A^*| \leq t$; contradiction. Hence $A = B$. Thus the polynomial $X^t = 1$ has t distinct roots in C, so is certainly separable (p. 57), and C is clearly a splitting field for $X^t - 1$ over K. Hence C is a Galois extension of K, by

Theorem 3.8. Let a be a generator of A, and τ a K-algebra automorphism of C. Then $A = \{x \in C : x^t = 1\}$ so $A = A\tau$, so $a\tau = a^{e(\tau)}$ for some integer $e(\tau)$ coprime to t and uniquely defined by a and τ. It is straightforward to check that the composite map $\tau \to e(\tau) \to U(\mathbb{Z}/t\mathbb{Z})$ is a group monomorphism. Thus we have proved (c) and (d).

Finally, suppose char $K = p \neq 0$ and K contains only finitely many roots of 1. If $T^* = \{y \in K : \exists\, r \geq 1, y^r = 1\}$ then T^* is a group, and by hypothesis is finite, hence cyclic. Let P be the prime field (p. 58) of K, and F be the algebra generated over P by T^*. Since $F \subseteq K$, it is clear that F is an integral domain. Since $\dim_P F \leq |T^*| < \infty$, F is finite. By Lemma 2.27, F is a division ring, and F is commutative (since K is), hence $F \simeq GF(p^f)$ for some $f \geq 1$, and $T^* = U(F)$, i.e. $F = T^* \cup \{0\}$. This proves (e)(i).

Now consider the subring F(A) of C generated by F and A. It is, by the same argument, a field of dimension $n \geq 1$ over F, and is a splitting field for $X^t - 1$ over F (since $X^t - 1$ is separable as already noted). It follows that $F[A] \simeq GF(p^{fn})$, where n is the least integer such that $t \mid (p^{fn} - 1)$, from page 59. It remains to show that $n = \dim_K C$. Let a be a generator of A, as in the proof of (c) and (d). Then $\dim_K C$ is the degree of the minimum polynomial Q(X) of a over K. But Q(X) splits into monic linear factors in F[A][X], so every monic factor in C[X] is in F[A][X] (using unique factorisation). It follows that $Q(X) \in F[X]$, and

$$n = \dim_F F[A] = \dim_K C = \deg Q(X) ,$$

as required. //

<u>4.29 Lemma.</u> Let K be a field, M an n-dimensional right K-vector space (where $n < \infty$), and G a group of K-automorphisms of M with abelian normal subgroup A. Suppose that R is the subring of $\mathrm{End}_K(M)$ generated by K and A, and that all the simple R- submodules of M are isomorphic, say to N where $\dim_K N = k$. Then

(a) $M \simeq N^s$ as R-module, where $s = n/k$ is an integer;

(b) R is a field containing K, and $\dim_K R = k$.

Furthermore, if A is finite then

(c) G is a subgroup of the group Γ of K-linear R-semilinear bijections of M (under the familiar convention that M is a right R-vector space of dimension s, by commutativity), and

(d) $C_G(A) = G \cap \mathrm{End}_R(M)$.

Note. The existence of a simple R-submodule of M follows from the fact that $\dim_K M < \infty$.

Proof. Clifford's Theorem yields (a) immediately, and (4.28) applied to N yields (b). Since R is a field it has one isomorphism class of simple module, so $N_i \simeq R \simeq N$ for each i. Hence $M \simeq R^s$. Clearly G is a group of K-linear bijections of M, so to obtain (c) we need to show that, given A is finite then each $g \in G$ is R-semilinear. Note that if A is finite then A is cyclic, with generator say a, and then $R = K[a]$. Hence $1, a, \dots, a^{k-1}$ is a K-basis of R. Furthermore, suppose $\sum_0^r p_i a^i = 0$ for some $p_i \in K$, and let $0 \neq w \in M$. Then

$$0 = w(\sum_0^r p_i a^i) = w(\sum_0^r a^i p_i),$$

so

$$0 = g0 = g\{w(\sum_0^r a^i p_i)\}$$

$$= \sum_0^r (g(wa^i))p_i$$

$$= \sum_0^r (g(a^i w))p_i$$

$$= \sum_0^r (ga^i g^{-1} gw)p_i$$

$$= \sum_0^r ((gag^{-1})^i gw)p_i$$

$$= \sum_0^r (a^{v(g)i} gw)\, p_i \quad \text{for some natural number } v(g) \text{ depending on } g$$

$$= \sum_0^r \{(gw)\, a^{v(g)i}\}\, p_i$$

$$= (gw)\, \{\sum_0^r a^{v(g)i} p_i\}.$$

Hence $\sum_0^r a^{v(g)i} p_i = 0$, and it follows that, for each $g \in G$, $g^{-1}ag$ has the same minimum polynomial over K as does a. Let $b \in R$ and $b = \sum_0^{h-1} b_i a^i$ (where the b_i are unique elements of K). Then it follows that the map $\tilde{g} : R \to R$ defined by $b\tilde{g} = \sum_0^{h-1} b_i (gag^{-1})^i$ is a K-algebra automorphism of R, and from the calculation just made it further follows that, if $x \in M$, then

$$g(xb) = g(x)\,(b\tilde{g}),$$

as required.

From the definition of $\tilde{g}$, it is clear that $C_G(A) = G \cap \mathrm{End}_R(M)$. //

The previous Lemma has the following interesting consequence.

4.30 Corollary. With the situation of (4.29), if $C_G(A)$ is finite then so is G, and $G/C_G(A)$ is isomorphic to a subgroup of the Galois group $\mathrm{Gal}(R/K)$, and hence has order at most k.

Proof. Since $A \subseteq C_G(A)$, we know that A is finite, and hence from the Lemma that $G/C_G(A)$ embeds in $\Gamma/\mathrm{Aut}_R(M) \simeq \mathrm{Gal}(R/K)$. We know that $|\mathrm{Gal}(R/K)| \leq k$ from (3.4) //

APPLICATIONS OF CHARACTER THEORY

In this section we use the character theory developed earlier to provide a criterion for non-simplicity of finite groups. We have however omitted one famous consequence of these results, Burnside's $p^a q^b$-theorem. This states that any group of order $p^a q^b$ where p and q are primes is soluble. It is straightforward deduction from the non-simplicity result of this section.

Our next result is a technicality required for the proof of Proposition 4.32.

4.31 Lemma. Let G be a finite group and $K \subseteq \mathbb{C}$ a splitting field for every subgroup of G. Let M be a simple KG module of degree n, and χ the character afforded by M. Suppose $g \in G$ and the conjugacy class of g has order coprime to n. Then $\chi(g) = 0$ or $|\chi(g)| = n$.

Proof. Let $\{1\} = C_1, \ldots, C_s$ be the distinct conjugacy classes of G. If $1 \leq j \leq s$, let $h_j = \sum_{x \in C_j} x$. From the proof of (4.12(b)), $h_1, \ldots, h_s$ is a K-basis of Z(KG). Since each product $h_i h_j \in Z(KG)$, there are elements $r_{ijk} \in K$ such that

$$h_i h_j = \sum_{k=1}^{s} r_{ijk} h_k \qquad 1 \leq i, j \leq s.$$

Clearly r_{ijk} is the number of pairs x, y where $x \in C_i$, $y \in C_j$ and $xy = z$ for some fixed $z \in C_k$ so each r_{ijk} is a non-negative integer. From the definition of a splitting field, and (4.7), we have

$$W(M) \simeq B(M) \simeq M_n(K)$$

where W(M) is the Wedderburn component of M in KG, B(M) is the bicommutator of M, and $M_n(K)$ is the ring of $n \times n$ matrices over K. For each h_j the element $\hat{h}_j$ of B(M) is central (since h_j is central in KG and B(M) is a homomorphic image of KG). Since $Z(M_n(K)) = K$, each h_j acts on M as scalar multiplication, say by $w_j \in K$. Thus $\chi(h_j) = \chi(1) w_j$, so by (4.11) $|C_j| \chi(x) = \chi(1) w_j$ for any $x \in C_j$, so

$$w_j = |C_j| \chi(x) n^{-1}.$$

From the defining relations of the r_{ijk}, we have, for any i, j, the equation

$$w_i w_j = \sum_{k=1}^{s} r_{ijk} w_k .$$

Fixing j, we obtain s simultaneous equations in $w_1, \ldots, w_k$:

$$\sum_{k=1}^{s} (r_{ijk} - \delta_{ik} w_j) w_k = 0 \qquad 1 \leq i \leq s .$$

Since $C_1 = \{1\}$ we have $h_1 = 1$ so $w_1 = 1$. Hence, for each j,

$$\det \begin{Bmatrix} r_{1j1} - w_j, & \ldots\ldots\ldots, & r_{1js} \\ r_{sj1}, & \ldots\ldots\ldots, & r_{sjs} - w_j \end{Bmatrix} = 0 ,$$

so w_j is an algebraic integer.

Suppose $g \in C_j$ and $|C_j| = h$, where $(h, n) = 1$. There exist rational integers u, v such that $uh + vn = 1$. But then

$$uh\chi(g)n^{-1} = (1 - vn)\chi(g)n^{-1}$$

is an algebraic integer, so $\chi(g)n^{-1}$ is. Suppose $\chi(g) \neq 0$, i.e. $B = \chi(g)n^{-1} \neq 0$. From page 62 the minimum polynomial of B over $\mathbb{Q}$ has integer coefficients; let it be

$$f(X) = X^m + a_1 X^{m-1} + \ldots + a_m .$$

If $B = B_1, \ldots, B_m$ are the distinct roots of $f(X)$ in some splitting field P of $f(X)$ over $\mathbb{Q}$, we have

$$f(X) = (X - B_1) \ldots (X - B_m)$$

in $P[X]$. Hence

$$a_m = B_1 \ldots B_m .$$

But $\chi(g)$ is a sum of n roots of 1, so $|B_i| \leq 1$, $1 \leq i \leq m$. On the other hand, $B \neq 0$ so $a_m \neq 0$. Hence $|a_m| = 1$, so $|B_i| = 1$, $1 \leq i \leq m$. Again because $\chi(g)$ is a sum of n roots of 1, they must all be the same to allow

$|B| = 1$. That is, g acts as scalar multiplication on M and $|\chi(g)| = n|B| = n$. //

4.32 Proposition. Let G be a finite group of composite order, and suppose $1 \neq g \in G$ and the conjugacy class of g has prime power order p^b for some prime p. Then G has a normal subgroup H such that $1 \neq H \neq G$; that is, G is not a simple group.

Proof. Let $K \subseteq \mathbb{C}$ be a field which is a splitting field for every subgroup H of G. If $b = 0$ then $Z(G) \neq 1$ and G is certainly not simple, so we can assume $b \geq 1$. Let $\chi_1, \ldots, \chi_s$ be the distinct dissimilar characters of G, with χ_1 the unique character such that $y\chi_1 = 1$ for every $y \in G$. Then (4.12)

$$|G| = \sum_1^s \chi_i(1)^2$$
$$= 1 + \sum_2^s \chi_i(1)^2 .$$

Now $p^b = [G : C_G(g)]$ so $p^b \,\Big|\, |G|$. Hence $p \nmid \chi_i(1)$ for at least one i, $2 \leq i \leq s$. Suppose that, when $2 \leq i \leq s$ and $p \nmid \chi_i(1)$ then $\chi_i(g) = 0$. Then from (4.13) and (4.14)

$$0 = \sum_1^s \chi_i(1)\chi_i(g)$$
$$= 1 + \sum_2^s \chi_i(1)\chi_i(g) .$$

If T is the ring of algebraic integers of K, it then follows that for some $w \in T$,

$$0 = 1 + pw$$

so $p^{-1} = -w$ is an algebraic integer: contradiction.

Hence we can find a simple KG module M, which affords the representation X and the character χ, and is of degree n, such that: $p \nmid n$, $\chi(g) \neq 0$, and for some $h \in G$ hX is not the identity on M. But by (4.31) $|\chi(g)| = n$, so that gX acts as scalar multiplication on M. Thus the subgroup $N = \{x \in G : xX$

acts as scalar multiplication on $M\}$ is a normal subgroup other than 1. Suppose $N = G$, then since M is simple we must have $\dim_K M = 1$. But then $X : G \to U(\mathrm{End}_K M)$ is a representation of G in the abelian group $U(\mathrm{End}_K M) \simeq U(K)$, and $hX \neq 1$ for some $h \in G$. It follows that G/G' has two dissimilar representations of degree 1, so $G \neq G'$. But since $p^b \neq 1$ G is not abelian, so $1 \neq G'$.

That is, either $1 \neq N \neq G$ and N is normal in G, or $1 \neq G' \neq G$; in either case, G is not simple. //

4.33 Lemma. Let G be a finite group with subgroups A, B such that A is cyclic, B is a p-group for some prime p, and $G = AB$. Then G is soluble.

Proof. If G is a p-group then G is nilpotent (1.16) hence soluble. So we may assume G is not a p-group. Let B^* be a Sylow p-subgroup of G containing B. Then $G = AB^*$, so we may assume $B = B^*$. Let $a \in A$, $a \neq 1$. Then $C_G(a) \geq A$ as A is abelian. Thus $[G : C_G(a)] \mid [G : A]$, and so either $C_G(a) = G$ or $[G : C_G(a)] = p^b$ for some $b \geq 1$. In either case, by (4.28) if necessary, G is not simple.

If $|A| = tp^w$, where $w \geq 0$, and $(t, p) = 1$ then A contains cyclic subgroups A_1, B_1 such that $|A_1| = t$, $|B_1| = p^w$, and then $G = A_1 B$. Hence we may assume $A = A_1$, and $|A| = t$. Since G is not simple there is a normal subgroup M such that $1 \neq M \neq G$. Using an inductive argument, we may assume that G/M is soluble, since clearly $G/M = (AM/M)(BM/M)$. Now we are assuming $|A|$ and $|B|$ are coprime, so we cannot have both $AM = G$ and $BM = G$. It is straightforward to check that, if $AM \neq G$ or $BM \neq G$ then AM (or BM) is a group satisfying the hypotheses of the theorem, and hence by an inductive argument is soluble. Every subgroup of a soluble group is soluble (Exercise!), so M is soluble. It then follows that G is soluble; the details are left to the reader. //

FINITE GROUPS

In this section we use representation and character theory to obtain several useful results about finite groups.

For the first we recall (p. 20) that an extraspecial group is a finite non-abelian p-group (for some prime p), such that $\Phi(G) = G' = Z(G)$, $|Z(G)| = p$ and $Z(G)$ is elementary abelian. Also $G/G' = G/\Phi(G)$ is elementary abelian too.

4.34 Proposition. Let G be an extraspecial p-group, and K a field which is a splitting field for every subgroup of G. Suppose either K is a subfield of the complex field or K is finite and char $K = q < \infty$ where $q \neq p$. Then

(a) $|G| = p^{2a+1}$ for some $a \geq 1$, and

(b) the group algebra KG is the direct sum of p^{2a} copies of K and $p - 1$ copies of $M_{p^a}(K)$; equivalently, KG has p^{2a} non-similar irreducible representations of degree 1 and $(p - 1)$ non-similar irreducible representations of degree p^a.

Proof. We only sketch the proof of (a). Let F be a field of p elementa, and regard $H = G/Z(G)$ as an F-vector space. From page 20 (but this time with the additive notation for H), let $(\, , \,) : H \times H \to F$ be the form discussed there. For any $x \in H$ let $x' \in G$ be an inverse image of x. If $0 \neq x \in H$ then $x' \notin Z(G)$, so there exists $y \in H$ such that $(x, y) = b \neq 0$, where $b \in F$. Let $z = b^{-1}y$; then $(x, z) = 1 = -(z, x)$, and of course $(w, w) = -(w, w) = 0$ for all $w \in H$. If $H = xF \oplus zF$ we are home. Otherwise, it is easy (using a similar trick which gave us z given x and y) to show that there is an F-subspace D of H such that $(x, d) = (y, d) = 0$ for every $d \in D$, and $H = xF \oplus zF \oplus D$. Then by an obvious inductive argument it follows that $\dim_F H$ is even, say 2a. Hence $|H| = p^{2a}$; since $|Z(G)| = p$ we have $|G| = p^{2a+1}$, as required.

For (b), assume first that $K \subseteq \mathbb{C}$. Every simple KH module is a simple KG module, so K is a splitting field for H. Since H is abelian of order p^{2a}, $KH \simeq K \oplus \dots \oplus K$ (p^{2a} copies), from Theorem 4.10, and KH is a homomorphic image of KG. But every ideal of KG is a direct summand, so $KG \simeq A \oplus KH$ for some semi-simple K-algebra A (where K is easily seen to be a splitting field for A).

Let $x \in G$, $x \notin G'$. The map $a \to [a, x]$ is a homomorphism of G onto G', from Lemma 1.10(b). The kernel is $C_G(x)$, so $[G : C_G(x)] = p$. Hence the conjugacy class of x has p elements. Since there are $p^{2a+1} - p$ elements of $G - G'$, there are $p^{2a} - 1$ conjugacy classes of order p. Since $|Z(G)| = p$, G has exactly p conjugacy classes of order 1, and hence $p^{2a} + (p - 1)$ conjugacy classes altogether. Hence by (4.12) G has exactly $p - 1$ irreducible representations of degree greater than 1.

Let $M_1, \ldots, M_{p-1}$ be the corresponding non-isomorphic simple modules, and $X_1, \ldots, X_{p-1}$ the representations and $\chi_1, \ldots, \chi_{p-1}$ the characters which they afford. Since K is a splitting field for G', Theorem 4.10 implies $KG' \simeq K \oplus \ldots \oplus K$ (p copies). Since G' is a cyclic group, say $G' = \langle g \rangle$, an adaptation of the proof of (4.17), and a counting argument, show that the distinct characters $\lambda_0, \lambda_1, \ldots, \lambda_{p-1}$ of G' are defined by $\lambda_i(g) = w^i$ for some primitive p^{th} root of 1, w, in $\mathbb{C}$.

Since $G' = Z(G)$ and $\chi_1, \ldots, \chi_{p-1}$ are irreducible characters over the splitting field K, $\chi_i|_{G'} = \chi_i(1)\lambda_j$ for some j, $1 \leq j \leq p - 1$. If χ is any character of G of degree 1 then $\chi|_{G'}$ is just $\chi(1)|_{G'}$. Thus using the scalar product notation $(\chi|_{G'}, \lambda_i) = 0$ for $1 \leq i \leq p - 1$. Hence for some e_{ij} in K

$$\lambda_i^G = \sum_{j=1}^{p-1} e_{ij}\chi_j, \qquad 1 \leq i \leq p - 1,$$

by Frobenius reciprocity. Since $\lambda_i^G(1) = p^{2a}$, not all e_{ij}'s are zero. Since $\chi_i|_{G'}$ is a multiple of exactly one λ_j, and the number of λ's is the same as the number of χ_i's we may after a suitable renumbering assume that

$$\lambda_i^G = \chi_i(1)\chi_i, \qquad 1 \leq i \leq p - 1.$$

Thus $\chi_i(1) = p^a$, and the result follows for $K \subseteq \mathbb{C}$.

The case of a finite field of suitable characteristic then follows from (4.27), and we leave it to the interested reader. The only difficult point is to show that, if $K \subseteq F$ are finite fields which are splitting fields for a finite group X, and if

$FX \simeq \bigoplus_{i=1}^{s} M_{n_i}(F)$ then $KX \simeq \bigoplus_{i=1}^{s} M_{n_i}(K)$. The details of this step resemble those of the final sections of the proof of (4.27). //

Suppose G is a group, C a commutative ring, M a (right) C-module and $X : CG \to End_C(M)$ is a representation of G. We say X is a <u>faithful</u> representation, or that G is <u>faithfully represented</u> on M, if the only element of G which acts as the identity on M is 1, i.e. if $G \to U(End_C(M))$ is an embedding. This is simply saying, in the terminology of Chapter 1 (page 2), that $X : G \to S^M$ is a faithful permutation representation.

It should be pointed out that the concepts of a faithful representation (over C) and a faithful C-module are quite distinct. For example, let $\mathbb{C}$ be the complex field and G a cyclic group of order 2, with generator g. Then the embedding $G \to U(\mathbb{C})$ gives a $\mathbb{C}G$ module structure on $\mathbb{C}$, so that $G \to U(\mathbb{C})$ is injective but $\mathbb{C}$ is not a faithful $\mathbb{C}G$ module (because its annihilator is the non-zero ideal $(1-g)\mathbb{C}G$).

For our purposes, faithful representations are important - by their absence - in the case of certain finite groups.

<u>4.35 Proposition.</u> Let A be an abelian group of prime power order p^n, and let M be a simple KA module, where K is a field of characteristic p. Then $bm = m \;\forall b \in A \;\; \forall m \in M$.

<u>Proof.</u> Let $0 \neq x \in M$. Then $M = KAx$. Suppose a is an element of A, and $ax \neq x$. Then $(1-a)x \neq 0$, so $M = KA(1-a)x$. Hence, for some $w \in KA$ $x = w(1-a)x = (1-a)wx$ (since KA is commutative). By an obvious inductive argument, $x = (1-a)^{p^q} w^{p^q} x$ for $q \geq 1$. But $0 = (1-a)^{p^r} = 1 - a^{p^r}$ for some $r \geq 1$, by the binomial theorem; contradiction. Hence $ax = x$. Clearly $a0 = 0$, so $bm = m \;\forall b \in A \;\forall m \in M$. //

The above proposition is used to establish the following important result.

4.36 Lemma. Let G be a finite group and K be a field of prime characteristic p. Suppose $N \neq 1$ is a normal p-subgroup of G, and M is a simple KG module. Then $G \to U(\mathrm{End}_K(M))$ is not a faithful representation of G.

Proof. If $C = Z(N)$ then $C \neq 1$ since N is a p-group and C is normal in G, since N is. By Clifford's Theorem M is a semisimple KC module. By (4.35), $cm = m \ \forall c \in C \ \forall m \in M$, since the same is true in each simple KC-submodule of M. //

REMARKS

1. With the definitions used in this book, the categories of rings and of ringlets (and the appropriate morphisms) are strikingly different. For example, among rings the inclusion of a non-trivial direct summand is not a morphism, while the "same" map (set-theoretically speaking) is a morphism in the category of ringlets. It is a non-trivial fact that a ringlet automorphism of a ringlet that happens to be a ring "is" (after appropriate categorical technicalities) a ring automorphism. With these problems in mind in this Chapter we have defined an algebra in such a way that inclusion of a non-trivial direct summand is a morphism of algebras (though not of unital algebras with their natural notion of morphisms).

2. It is of interest from a ring-theoretic viewpoint that there are fields K and groups H, G (where H is a normal subgroup of G) such that some simple KG module is not a semisimple KH module. For example, let $\mathbb{Q}$ be the rational field and P the pure transcendental field extension of $\mathbb{Q}$ by one central indeterminate t. Let $G = U(P)$ and H be the cyclic multiplicative subgroup of G generated by t. Then the interested reader will be able to check that P is naturally a simple $\mathbb{Q}G$ module, but has no simple $\mathbb{Q}H$ submodules (under restriction of scalars).

3. We have not had cause to use the term in this Chapter, but in the next one we shall need the concept of an irreducible group. Suppose G is a group, K a field, KG the group algebra, and M a KG module. Then a subgroup H is irreducible (for M) if M is a simple KH module (and, consequently, also a

simple KG module). Similarly, when the context permits we occasionally speak of a G-module (where G is a group), rather than a KG-module (where K is a commutative ring); in such situations, the identity of the relevant ring K will be clear from the context.

5 Hering's theorem and applications

This final chapter is in three sections. The first deals with the proof of Hering's results mentioned in the title. These are quite complex and involve results from throughout the preceding chapters. The middle section is a brief introduction to Block Designs, in which we develop the subject from a fairly abstract standpoint. The final section applies Hering's results to Block designs and to 2-transitive groups (as defined therein).

HERING'S RESULTS

We begin by recalling a definition given earlier. If n and q are natural numbers greater than 1, let $\Phi_n(x)$ be the nth cyclotomic polynomial and define $\Phi_n^*(q) = \Phi_n(q)/e^d$ where $e = \text{h.c.f.}(n, \Phi_n(q))$ and e^d is the highest power of e dividing $\Phi_n(q)$ (see 3.23). It follows from (3.23) that $\Phi_n^*(q)$ and n are coprime. We can now prove a proposition which relates various properties of a prime r and the irreducible representations of r-groups.

5.1 Proposition. Let $q \geq 2$ be a prime power and let $n \geq 1$ be a natural number. If r is a prime number then the following conditions on r, n, q are equivalent:

(a) $r \mid \Phi_n^*(q)$;

(b) $r \nmid q$ and q has multiplicative order n (mod r);

(c) $GL(n, q)$ contains non-trivial r-subgroups and all such subgroups are irreducible (for an n-dimensional space over $GF(q)$);

(d) $GF(q^n)$ contains an element of multiplicative order r which does not lie in any proper subfield that contains $GF(q)$.

Proof. Before starting the proof proper, we make two comments. First, let $GL(n, q)$ be the general linear group of dimension n over the finite field $GF(q)$ of q elements; $GL(n, q)$ is the group of automorphisms of an n-dimensional (right) $GF(q)$ vector space V. Then

$$|GL(n, q)| = (q^n - 1)(q^n - q) \ldots (q^n - q^{n-1}),$$

because, given any specific ordered basis $\mathfrak{X}$ of V, there is a natural correspondence between the elements of $GL(n, q)$ and the images of $\mathfrak{X}$ under these elements.

Let a, z be coprime integers, and suppose a has multiplicative order $b \bmod z$. Since $|U(\mathbb{Z}/z\mathbb{Z})| = \phi(z)$, we know $b | \phi(z)$. (a) implies (b). Since $r | \Phi_n^*(q)$ we have $r | \Phi_n(q)$, so $r | (q^n - 1)$. Thus $(r, q) = 1$ so $r \nmid q$. By (3.21) $n = fr^i$ where f is the order of $q \pmod r$, and $i \geq 0$. If $i > 0$ then $r | n$ and hence by the definition of $\Phi_n^*(q)$ (3.23) $r \nmid \Phi_n^*(q)$. Hence $i = 0$, and n is the order of $q \pmod r$.

(b) implies (c). Since q has order $n \pmod r$, r divides $q^n - 1$. Since $q^n - 1$ divides the order of $GL(n, q)$ by the remark above, r divides the order of $GL(n, q)$. Thus $GL(n, q)$ contains non-trivial r-subgroups. Let R be an r-subgroup of $GL(n, q)$. Now V is the underlying vector space of dimension n over $GF(q)$, so let W be an R-invariant subspace of V of dimension m. If $m < n$, we have r divides the order of $GL(m, q)$. But

$$|GL(n, q)| = q^w \prod_{t \leq m} (q^t - 1) \quad \text{for some } w.$$

Thus $r | (q^t - 1)$ for some $t < n$ and so n is not the order of $q \pmod r$; contradiction. Hence $m = n$ and V has no proper R-invariant subspace.

(c) implies (d). Certainly $GL(n, q)$ contains a cyclic subgroup of order r, and the result then follows from Theorem 4.28.

(d) implies (a). Since the multiplicative group of a field is cyclic, $GF(q^d)$ contains an element of order r if and only if $q^d \equiv 1 \pmod r$. Under the conditions of (d) it follows that r is a primitive prime divisor of $q^n - 1$ (see page 72), so that q has multiplicative order $n \pmod r$ and $r | \Phi_n(q)$. But $\phi(r) = r - 1$ so $n | (r - 1)$, hence $(n, r) = 1$. From the definition of $\Phi_n^*(q)$ (3.23), it follows that $r | \Phi_n^*(q)$. //

For the remainder of this section we shall fix the following notation. Let K be a finite field of order q and characteristic p, and V be a right K-vector space of dimension n. Let G be a subgroup of GL(n, q) such that $(|G|, \phi_n^*(q)) \neq 1$. Clearly this implies that there is a prime r such that $r | \phi_n^*(q)$ and G has a non-trivial r-subgroup. If P is any such non-trivial r-subgroup then V is a simple left KP module by (5.1). Let R be a Sylow r-subgroup of G, and S the normal closure in G of R. Also, let L be the centraliser of KS in $\mathrm{End}_K(V)$. Since V is a simple left KS module L is a division ring, and hence by Wedderburn's theorem a (commutative) field. Let $\dim_K L = m$, so that $|L| = q^m$, and $\dim_L V = n/m$. Since L is commutative V can be thought of as an L - L bimodule in the way discussed in Chapters 2 and 4, so V is a left LG module. Finally, let F be the Fitting subgroup of G.

<u>5.2 Lemma.</u> Let A be an Abelian subgroup of G normalised by R. Then A is cyclic and centralised by R. In particular if A is normal in G then $A \leq C_G(S)$.

<u>Proof.</u> We observe that V is a simple KAR module. It is also clear from pages 58 and 59 that, for any prime power q and any d, the group $\mathrm{Gal}(GF(q^d)/GF(q))$ is cyclic of order d, generated by the map $x \mapsto x^q$ (itself a power of the Frobenius automorphism of $GF(q^d)$). It then follows from (4.24) that A is cyclic and $[N_G(A) : C_G(A)] \mid n$. From the proof of (5.1), $r \equiv 1 \pmod n$, so $R \subseteq C_G(A)$. If A is normal in G then so is $C_G(A)$, so $S \subseteq C_G(A)$. //

<u>5.3 Lemma.</u> R and $C_G(R)$ are cyclic groups (so $R \subseteq C_G(R)$) and $|C_G(R)| \mid (q^n - 1)$. Furthermore, $[N_G(R) : C_G(R)] \mid n$.

<u>Proof.</u> Let R_0 be a subgroup of Z(R) of order r. Since $R \subseteq C_G(R_0)$ it will suffice to consider the case of R_0. In that case the result follows from (4.29), (4.30) and the observation in the proof of (5.2) that, for a prime power p and natural number d, $\mathrm{Gal}(GF(q^d)/GF(q))$ is cyclic of order d. //

5.4 Theorem. Assume that F is not centralized by S. Then the following statements are true:

(a) $|R| = r = n + 1 = 2^a + 1$ for some $a \in \mathbb{Z}$, $a > 0$;

(b) $|L| = q$ and q is odd;

(c) there is a normal subgroup N of G such that

(i) $F = NZ(G)$, and

(ii) $|Z(N)| = 2$ or 4,

(iii) there is an extraspecial 2-group T of order 2^{2a+1} such that $N = TZ(N)$,

(iv) G/F is faithfully represented on $N/Z(N)$;

(d) $C_{G/F}(RF/F) = RF/F$, $C_F(R) = Z(G)$ and $F(G/F) = RF/F$.

Proof. Since $[F, S] \neq 1$ and F is a normal subgroup, an elementary calculation shows that $[F, R] \neq 1$. There is a prime π such that F contains a π-element, and hence a Sylow π-subgroup P, which does not centralise R. By (5.3) R is cyclic and so $[R, R] = 1$. Thus $\pi \neq r$. Since V is a simple PR-module and P is normal in PR, $\pi \neq p$ by (4.28).

Let N be a subgroup of G minimal with respect to satisfying the three conditions: N is normal in G, N is a subgroup of P, and $[N, R] \neq 1$. If $N \cap S \neq N$ then $[N \cap S, R] = 1$. But $[N, R] \subseteq N \cap S$, so then $[N, R, R] = 1$. Since $(r, \pi) = 1$ (1.26) shows that $[N, R] = 1$, which contradicts the choice of N. Hence $N = N \cap S$ so $N \subseteq S$.

Now V is a simple KR-module and P is a normal subgroup of PR, while the elements of S commute with those of L. Hence N has a faithful representation (p. 118) over L. Now N is not abelian, since otherwise $[N, R] = 1$ by (5.2). Let Y be a subgroup of N such that N/Y is a chief factor of G. Then by the minimality of N, $Y \subseteq C_G(R)$.

By (5.3) Y is cyclic and $Y \leq C_N(Y)$. Since Y is normal in G and $Y \subset N \subseteq P$, we must have $[Y, R] = 1$. But $(|Y|, |R|) = 1$, so $YR = Y \times R$, and so $YR \leq C_G(Y)$. Assume that $Y = C_N(Y)$ then $YR = C_{NR}(Y)$. But $C_{NR}(Y)$ is normal in NR and so YR is normal in NR. But R is a normal Sylow r-subgroup of YR and so R is characteristic in YR. This implies R is normal in NR. However R and P are Sylow subgroups of NR and so

$[N, R] \leq N \cap R = 1$. But this contradicts the choice of N. Thus $Y < C_N(Y)$ and so $C_N(Y) = N$ as N/Y is a chief-factor of G.

Thus $Y = Z(N)$ since $Z(N) = N$ would imply that N is Abelian. Let X be a proper characteristic subgroup of N. Then X is normal in G. Let X_0 be a subgroup of N such that $X_0 \geq X$ and N/X_0 is a chief-factor of G. Then by the above arguments $X_0 = Z(N) = Y$. So $X \leq Y$. Hence every proper characteristic subgroup of N is contained in Y and is thus central and cyclic. Since N is non-Abelian, by (1.31) N is extraspecial or $\pi = 2$ and $N = TZ(N)$ where T is an extraspecial group of order at least 16 and $Z(N)$ is cyclic of order 4.

In either case let T denote the extraspecial group. Let T have order π^{2a+1}. Since $T \leq N \leq S$, V is naturally a left LT module, and $\dim_L V = n/m$ where $m = [L : K]$. Furthermore, LT is semisimple by Maschke's Theorem, and it is clear that at least one simple LT-submodule W of V has L-dimension greater than 1. It follows from (4.31) that $n/m \geq \pi^a$.

We now consider the action of R on N. Let $C = C_N(R)$. By (5.3) C is cyclic. Since $Y \leq C_G(R)$, $Y \leq C$. Let $D = C_N(C) \geq C \geq Y$. If $C \neq Y$, $D < N$ since $D \neq Z(N)$. So suppose $D \neq N$. Since N/Y is an elementary abelian π-group, C is normal in N, and so N/D is a non-trivial subgroup of the automorphism group of C.

Now if $\pi \neq 2$ then Aut(C) is cyclic, so N/D is cyclic. If $\pi = 2$, then N/D is cyclic of order 2 or $|N/D| \geq 4$. But if $|\mathrm{Aut}(C)| \geq 4$ then Aut(C) is the direct product of cyclic groups of orders 2 and 2^b for some $b \geq 1$, so $|N/D| = 4$ since N/D is elementary abelian. We will deal later with the case $|N/D| = 4$; suppose for the time being that N/D is a non-trivial cyclic group.

Since N/Y is an elementary abelian π-group and $(r, \pi) = 1$, and since D is R-invariant, there is a subgroup $E \geq Y$ of N such that

$$N/Y = D/Y \times E/Y$$

as $GF(\pi)R$-modules (by Maschke's Theorem (4.1)). Since $|N/D| \neq 1$ E/Y is cyclic and non-trivial. But $Y \leq Z(E)$ since $Y = Z(N)$, so E is abelian. Now

$C_E(R) = Y$ since $C \leq D$ and $[Y, R] = 1$, so by (1.30) $E = Y \times E'$ where $E' = [E, R] \neq 1$. On the other hand, if E is abelian and E is normalised by R then E is cyclic by (5.2), so E has no non-trivial direct decomposition; contradiction.

Thus, if $D \neq N$ we must have $|N/D| = 4$. But then, by similar arguments, there exist subgroups $D^{\#} \geq Y$ and $E^{\#} \geq Y$ of N such that

$$N/Y = C/Y \times D^{\#}/Y \times E^{\#}/Y,$$

as $GF(\pi)R$-module. But R has a homomorphic image which is a group of automorphisms of N/D, and $|Aut(N/D)| = 6$. Hence r must be 3, and it follows easily (compare p. 93) that, for a central subgroup R_0 of R, of order 3, every simple $GF(\pi)R_0$-module is of dimension 1 or 2 over $GF(\pi)$. (Recall that $\pi \neq r = 3$.) Since $C_N(R) = C$ and $Y \leq C$ and C/Y is cyclic, $\dim_{GF(\pi)}(D^{\#}/Y)$ and $\dim_{GF(\pi)}(E^{\#}/Y)$ are even and $\dim_{GF(\pi)}(C/Y)$ is 1. But $|N/Y| = \pi^{2a}$, since (as is easily seen from (1.31)) $N/Y \simeq T/T'$, so

$$\dim_{GF(\pi)}(C/Y) = \dim_{GF(\pi)}\{(N/Y)/(D^{\#}/Y \times E^{\#}/Y)\},$$

which is even; contradiction.

Thus we may assume $D = N$ and $C = Y$. Hence R fixes only the identity element of the group N/Y, since $[R, N] = N$ by (1.30) and (1.28). So $|R| \,\big|\, (|N/Y| - 1)$, and then $|R| \,\big|\, (\pi^{2a} - 1)$. But r, however, is a primitive prime divisor of $q^n - 1$, so $r \geq n + 1$. Thus

$$n + 1 \leq r \leq \pi^a + 1 \leq (n/m) + 1 .$$

This forces $m = 1$ and $r = \pi^a + 1$. Since r is prime $\pi = 2$, and then $r = 2^a + 1 = n + 1$. Both (a) and (b) now follow immediately.

We now need to establish (c). The existence of N and T has been shown, but it remains to prove c(i) and c(iv). Since $n = 2^a$ it follows easily from (4.34) that V is a simple KT module and that $\mathrm{End}_{KT} V \simeq K$. Hence $C_G(T)$ consists only of scalar multiplications, so $C_G(T) = Z(G)$. If $x \neq 2$ is prime then any Sylow x-subgroup of F will centralise T, by (1.16), so $F = UZ(G)$

where U is the Sylow 2-subgroup of F.

Let $Z(N) \leq X \leq G$ and $X/Z(N) = C_{G/Z(N)}(N/Z(N))$. Then $Z(G)N \leq X$. From (1.31) it can be seen that $N/Z(N) \cong T/Z(T)$ and so $N/Z(N)$ has order 2^{2a}. As $N/Z(N)$ is elementary abelian there are subgroups $N_1, \ldots, N_{2a}$ of N such that for each i $N_i \geq Z(N)$ and $[N : N_i] = 2$. As X centralises $N/Z(N)$, X acts trivially on $N_i/Z(N)$ but also X acts trivially on Z(N) since $Z(N) \leq Z(G)$. A simple argument similar to (1.27) shows that X induces an automorphism of order 2 on each N_i. Thus $[X : C_X(N_i)] = 2$ and

$$[X : Z(G)] = [X : C_G(F)] \leq [X : C_X(N)] \leq 2^{2a}.$$

But $N \leq X$ and $[N : C_N(N)] = 2^{2a}$ and so $X = NZ(G)$ as claimed.

This shows that $G/NZ(G)$ acts faithfully on $N/Z(N)$. But $N/Z(N)$ is an elementary abelian 2-group, and $N/Z(N)$ is a simple GF(2) R-module. From (4.36) $G/NZ(G)$ contains no normal Sylow 2-subgroup.

Thus if U is the Sylow 2-subgroup of F the $U \leq N(Z(G))$ so since $F = UZ(G)$ we have $F = NZ(G)$. This completes the proof of c(i) and c(iv).

We have already shown $r = 2^a + 1$ and so $r \mid \phi^*_{2a}(2)$. Since G/F is faithfully represented on $N/Z(N)$ we can apply the results of (a), (b) and (c) to G/F in its action on $N/Z(N)$. Since the characteristic of the field in this case is 2, (a), (b) and (c) cannot hold. Thus if F_1/F is the Fitting subgroup of G/F, $[R, F_1] \leq F$. Let $\bar{\ }$ denote the natural homomorphism from G to G/F. Recall that $N/Z(N)$ is a simple $GF(2)\bar{R}$-module and so by (4.28), $C_{\bar{G}}(\bar{R})$ has order dividing $2^a + 1$. But $r = 2^a + 1$ and so $C_{\bar{G}}(\bar{R}) = \bar{R}$. Hence $\bar{F}_1 = \bar{R}$.

Finally, as $F = NZ(G)$ and $C_N(R) = N \cap Z(G)$ we have $C_F(R) = Z(G)$, as required. //

<u>5.5 Theorem.</u> Let U be a subgroup of G so that $R \leq N_G(U)$ but $R \not\leq U$. Then U is nilpotent.

<u>Proof.</u> Let W be any Sylow w-subgroup of U for a prime w dividing $|U|$. By the Frattini argument $N_{UR}(W)U = UR$, so there is a conjugate of R normalizing

W. Hence there is a conjugate of W, also contained in U, which is normalized by R. Thus we may suppose that R normalizes W. Now WR is soluble as $W \lhd WR$ so $C_{WR}(F(WR)) \subseteq F(WR)$. By the previous theorem either $[F(WR), R] = 1$ or $w = 2$ and $r = n + 1 = 2^a + 1$. If $[F(WR), R] = 1$, then $R \subseteq F(WR)$ and $[R, W] = 1$. Then W is cyclic by (5. 3), and also from (5. 3) $C_U(R)$ is cyclic.

By the argument above $C_U(R)$ contains a Sylow subgroup of U for each odd prime dividing $|U|$. Either $C_U(R) = U$ and U is cyclic or $U = C_U(R)T$ where T is a Sylow 2-subgroup of U, $r = n + 1 = 2^a + 1$ and T contains a normal extraspecial group of order 2^{2a+1}. However by (4. 33) we know that U is soluble and so UR is also soluble. Let $H = F(UR)$. If $[H, R] = 1$ we have $R \subseteq H$ (1. 19) and so U is cyclic since $[R, U] = 1$. Thus $[U, R] \neq 1$ and we can apply (5. 4). We have that $H \leq U$ and so $F(UR/H) = RH/H$. Then $[R, U] \leq H$ and so $U/H \leq C_{UH/H}(R)$ but we already know that $C_{UH/H}(HR) = RH/H$ and so $U = H$ and hence U is nilpotent. //

5. 6 Theorem. Suppose K, q, V, G, R, S, F, L have the meanings defined before Lemma 5. 2. Then

(a) SF/F is simple;

(b) G/S is isomorphic to a quotient of the group of semilinear transformations of dimension 1 over $GF(q^n)$;

(c) $C_G(S) = Z(F)$ and Z(F) is a subgroup of the multiplicative group of the field L;

(d) if $|L| < q^n$ and d is a prime divisor of $(\phi_n^*(q), |G|)$, let D be a Sylow d-subgroup of G. Then $|D| \,\big|\, |SF/F|$;

(e) at least one of the following holds:

 (i) $|L| = q^n$;

 (ii) $|SF/F| = r$ and $C_G(SF/F) = SF/F$;

 (iii) $C_{G/F}(SF/F) = 1$;

(f) $F \cap S = F(S)$ is the unique maximal normal subgroup of S;

(g) if G is not soluble then

 (i) $S' = S$, and

(ii) G/SF is isomorphic to a subgroup of the outer automorphism group of SF/F, and

(iii) if $F \leq C_G(S)$ then $S \cap F \leq S' \cap Z(S)$.

Proof. (a) Suppose $F < X < SF$ where X is a normal subgroup of G. As $X < SF$, $R \not\leq X$. Hence by (5.5) X is nilpotent. But then $X \leq F$. So SF/F is a chief-factor of G. Hence it is a direct product of isomorphic simple groups. Since $R \leq SF$ and $R \not\leq F$, and R is cyclic, SF/F must therefore be simple.

(b) Since R is a Sylow subgroup of S, $G = N_G(R)S$ by the Frattini argument. Thus $G/S \cong SN_G(R)/S \cong N_G(R)/N_S(R)$. The result follows from (4.30).

(c) As $S \geq R$, $C_G(S) \leq C_G(R)$ thus $C_G(S)$ is cyclic by (5.3). Thus $C_G(S) \leq F$. If $F \subseteq C_G(S)$ then (c) follows at once. So we may assume that $C_G(S) < F$. We can now apply the results of (5.4) and deduce that $F = NZ(G)$ where $Z(N) \leq Z(G)$ and $N/Z(N)$ is a simple R-module. Hence $C_G(S) = Z(G) = Z(F)$ and clearly $Z(F)$ is a subgroup of the group of units of L.

(d) Assume that $|L| < q^n$. If there exists an abelian normal subgroup A of G such that $1 \neq A$ and V is a simple A-module then by (4.28) $|L| = q^n$; contradiction.

Note that the results (5.1) through (5.5) apply to any prime divisor d of $(\phi_n^*(q), |G|)$. Let D be a Sylow d-subgroup of G. Then D is cyclic by (5.3). If $D \not\leq S$, by (5.5) S is nilpotent, but then R is a normal abelian subgroup such that V is a simple R-module. Thus $D \leq S$. Furthermore, $D \cap F = 1$ since otherwise $D \cap F$ would be an abelian normal subgroup for which V is a simple $R_1 \cap F$-module. The result then follows.

(e) Assume (iii) does not hold. Then $C_{G/F}(SF/F) > F/F$. Let $U \geq F$ be a subgroup such that $U/F = C_{G/F}(SF/F)$. U is not nilpotent and so $U \geq R$ and hence $U \geq S$. But then SF/F is abelian and so $SF = RF$. If $R \leq F$ then as noted earlier $q^n = |L|$.

If $R \subsetneq F$, $[R, F] \neq 1$ and by (5.4(d)) $RF = SF$ and $C_{G/F}(RF/F) = RF/F$ and (ii) holds.

(f) If $X \neq S$ is a normal subgroup of S then $R \not\subseteq X$ and so X is nilpotent and $X \leq F$. So (f) follows immediately.

(g) Suppose G is not soluble. From (b) it follows that G/S is soluble. If $S' < S$ it would imply that $S' = F \cap S$ since $S' \not\geq R$ and so G would be soluble; contradiction. Hence (g) (i) holds.

Since G is not soluble $|L| < q^n$ and so $(G/F)/SF/F$ is isomorphic to a subgroup of the outer automorphism group of the simple group SF/F, which is g(ii), because of e(iii).

Finally if $F \leq C_G(S)$ we have that $S \cap F \leq S' \cap Z(S)$ as claimed. //

We will not directly use this major result but its importance is that it proves that (under the given hypotheses) either G is soluble or G has at most one non-abelian chief-factor which is simple. We refer to this later in this chapter. But first we deal with the final result of this section.

<u>5.7 Theorem.</u> Let G be a soluble subgroup of $GL(n, q)$ where q is a prime power. Assume that for some d such that $0 < d < n$ $\{(q^n - 1)/(q^d - 1)\}$ divides the order of G. Then at least one of the following holds:

(a) $n = 6$ and $q = 2$;

(b) $n = 2$ and $q + 1 = 2^j$ for some $j > 0$;

(c) G is a subgroup of $\Gamma L(1, q^n)$;

(d) $n = 4$ and $q = 3$;

(e) $n = 2$ and $q = 5$, 11 or 23.

<u>Proof.</u> Because $\phi_n(q)$ divides $\{(q^n - 1)/(q^d - 1)\}$ and $\{(q^n - 1)/(q^d - 1)\}$ divides $|G|$ we have that $\phi_n^*(q)$ divides $|G|$. We now have two distinct possibilities. Either $\phi_n^*(q) = 1$ or $\phi_n^*(q) > 1$.

If $\phi_n^*(q) = 1$ then either $n = 6$ and $q = 2$ or $n = 2$ and $q + 1 = 2^j$ for some j, by (3.24). These are the cases (a) and (b) respectively.

So assume $\phi_n^*(q) > 1$. Let r be a prime divisor of $\phi_n^*(q)$ and note that r divides $|G|$. Let F be the Fitting subgroup of G and let R be a Sylow r-subgroup of G. From (5.3) we know that R is cyclic. Assume $[R, F] = 1$ then $R \leq F$ since G is soluble, by (1.19). In this case R is normal in G, since R is characteristic in F. Thus $G \leq \Gamma L(1, q^n)$ by (4.30). Thus (c) holds.

We may thus assume that $[R, F] \neq 1$. Hence the conclusions of (5.4) hold. As a consequence

$$r = n + 1 = 2^n + 1 .$$

We also know from (5.4(d)) that $F = NZ(G)$ and G/F is faithfully represented on $N/Z(N)$ which is elementary abelian of order 2^{2a}. Thus G/F is a subgroup of $GL(2a, 2)$. Thus $|R| \,\big|\, |GL(2a, 2)|$ and since $r = 2^a + 1$ and $|GL(2a, 2)| = (2^a + 1)(2^a - 1)(2^{2a} - 2) \ldots (2^{2a} - 2^{2a-1})$ we deduce that $|R| = r$. Thus r divides $\phi_n^*(q)$ but $r^2 \nmid \phi_n^*(q)$.

Similar arguments apply to any primes dividing $\phi_n^*(q)$ and so if (c) does not hold then

$$s = n + 1 = 2^a + 1 = r .$$

Thus $\phi_n^*(q) = r$. Hence by (3.26) either $n = 4$ and $q = 3$ or $n = 2$. If $n = 4$ and $q = 3$ (d) holds so we may now suppose $n = 2$. In that case $r = 3$ and $a = 1$. However in this case G/F has order dividing $GL(2, 2)$, which is 12, and $N/Z(N)$ has order 4. But from (5.4(d)) the structure of N involves an extraspecial 2-group T such that $N = TZ(N)$ or $N = T$. Hence $2 \nmid q$. Also there is a well-behaved $GF(2)$ form on T/T' (see page 20), which must be preserved by the action of G on T, since $Z(T) \leq Z(G)$. It follows that G/F is a subgroup of the special linear group $SL(2, 2)$. Thus $|G/F|$ divides 6. So $|G/Z(G)|$ divides 24.

Now $(q^2 - 1)/(q^d - 1)$ divides $|G|$ and so $q + 1$ divides $|G|$. Since $\phi_2^*(q) = 3$ we see that $Z(G)$ has to be scalars and so $q + 1$ divides $|G/Z(G)|$. Thus $(q + 1) \,|\, 24$ and $q + 1 = 3.2^t$ for some t, $1 \leq t \leq 3$. Thus $q = 5$, 11 or 23, proving (e). //

DESIGNS AND STRUCTURES

The theorems we have just proved are results in linear algebra and group theory. The original purpose of them was to help study certain types of permutation groups. It was then realised that the techniques also had applications to automorphism groups of finite geometries. In this section we define the basic ideas of an incidence structure which we need in order to understand those applications.

The results proved are only those needed in the applications and we do not attempt a comprehensive coverage. There is considerable confusion about the language of block-designs and finite geometries. We have decided to follow the terminology of Hughes and Piper [15] in the hope that this may become standard.

For our proofs we adopt a slightly more abstract approach because this, it seems to us, gives more structure to the results. The notation comes from the history of the subject in statistical design theory so that the standard symbols are v (varieties), b (blocks), and r (replications).

However to conform with our usage in Chapter 1 points will be denoted by small Greek letters. A <u>structure</u> is a triple $(\Omega, \mathcal{B}, I)$ where Ω is the set of points, $\mathcal{B}$ is the set of blocks, $\Omega \cap \mathcal{B} = \emptyset$ and I is a relation on $\Omega \times \mathcal{B}$, i.e. a (non-empty) subset of $\Omega \times \mathcal{B}$. We usually write $|\Omega| = v$, $|\mathcal{B}| = b$ and if $(\alpha, B) \in I$ we will often denote this by $\alpha I B$ and write "α is incident with B".

The concept of a structure, in its full generality, is of little value but it has a number of interesting specialisations. Given integers t, k and λ with $t \geq 1$ and $\lambda \geq 1$ we say that a structure is a <u>$t-(v, k, \lambda)$</u> structure if $v < \infty$ and

(i) $\forall B \in \mathcal{B} \quad |\{\alpha | (\alpha, B) \in I\}| = k$ and

(ii) for any subset of t distinct points $\{\alpha_1, \ldots, \alpha_t\}$, $\alpha_i \in \Omega$

$$|\{B | (\alpha_i, B) \in I, 1 \leq i \leq t\}| = \lambda .$$

This is equivalent to saying that each block has k points incident with it and that every subset of t points is incident with exactly λ common blocks, and to avoid triviality we assume $k < v$. For example we can now list a $2-(6, 3, 2)$ structure by showing the set of points $\{0, \ldots, 5\}$ incident with each block $B_0, \ldots, B_9$.

$B_0 : 0$	1	2		$B_5 : 1$	2	5
$B_1 : 0$	1	3		$B_6 : 1$	3	4
$B_2 : 0$	2	4		$B_7 : 1$	4	5
$B_3 : 0$	3	5		$B_8 : 2$	3	4
$B_4 : 0$	4	5		$B_9 : 2$	3	5

We will denote this structure by $\mathcal{S}_1$.

After a quick count it is noticed that each point is incident with precisely 5 blocks. In fact such regularity always occurs.

5.8 Lemma. Let $(\Omega, \mathcal{B}, I)$ be a $t-(v, k, \lambda)$ structure and let s be an integer such that $0 \leq s < t$. Let $\Delta \subset \Omega$ such that $|\Delta| = s$. If λ_s is the number of blocks B such that every element of Δ is incident with B then $\lambda_s = \lambda \frac{(v-s)(v-s-1) \ldots (v-t+1)}{(k-s)(k-s-1) \ldots (k-t+1)}$.

Proof. We will write "ΔIB" or "Δ is incident with B" to mean $B \in \mathcal{B}$ and αIB for all $\alpha \in \Delta$. If B is a block such that ΔIB there are $\binom{k-s}{t-s}$ subsets of points incident with B, which contain Δ and have precisely t elements. Thus we obtain $\binom{k-s}{t-s}\lambda_s$ pairs (Δ_0, B) where $|\Delta_0| = t$, $\Delta_0 \supset \Delta$ and $\Delta_0 IB$.

However each t-subset (i.e. subset of exactly t points) of Ω is incident with λ blocks. There are $\binom{v-s}{k-s}$ subsets Δ_0 of Ω which contain Δ and each is incident with λ blocks, as just noted. So

$$\lambda_s\binom{k-s}{t-s} = \lambda\binom{v-s}{t-s}.$$

This is the required formula. //

Traditionally, λ_1 is denoted by r, a convention we shall also follow from now on. Since $k < v$, (5.8) implies that $r > \lambda$.

5.9 Corollary. If $(\Omega, \mathcal{B}, I)$ is a $t-(v, k, \lambda)$ structure then

(a) $b = \lambda \frac{v(v-1) \ldots (v-t+1)}{k(k-1) \ldots (k-t+1)}$

(b) if $t \geq 1$ then $bk = vr$,

(c) if $t \geq 2$ then $r(k-1) = \lambda_2(v-1)$. //

In most of the applications $t = 2$ and so $\lambda_2 = \lambda$.

Given the structure $\mathcal{S}_1 = \{\Omega_1, \mathcal{B}_1, I_1\}$ we can construct a new structure $\mathcal{S}_2 = \{\Omega_2, \mathcal{B}_2, I_2\}$ as follows: $\Omega_2 = \Omega_1$, $\mathcal{B}_2$ contains 20 blocks B_{ij}, $0 \leq i \leq 9$, $0 \leq j \leq 1$, and incidence is defined by: $\alpha I_2 B_{ij}$ if and only if $\alpha I_1 B_i$ (for $\alpha \in \Omega_1 = \Omega_2$ and $B_i \in \mathcal{B}_1$).

This new structure is a $2-(6, 3, 4)$ structure but has the property that the set of points incident with a block does not uniquely determine that block. A design or block-design or a t-design or $t-(v, k, \lambda)$ design is a $t-(v, k, \lambda)$ structure in which a block is determined by the set of points with which it is incident. Which of these terms is used will depend on the amount of information being given, though usually we will refer to a $t-(v, k, \lambda)$ design.

In this situation the blocks can be thought of as subsets of Ω by identifying the block with the set of points with which it is incident. As a consequence of (5.9) given t, v, k and λ for a block-design we know b and r.

We will now prove a result called Fisher's Inequality, which states that $v \leq b$. Before proceeding to the proof we will introduce some notation and ideas which will be used both in this proof and a later one. Given a structure $(\Omega, \mathcal{B}, I)$ and a field K we will let V_Ω be the K-vector space of all mappings from Ω to K and $V_{\mathcal{B}}$ be the K-vector space of all mappings from $\mathcal{B}$ to K.

Using the incidence relation I we can define two mappings T and S as follows:

$$T : V_\Omega \to V_{\mathcal{B}} \text{ is given by } B(fT) = \sum_{\substack{\alpha \in \Omega \\ (\alpha, B) \in I}} \alpha f \qquad \text{where } B \in \mathcal{B},\ f \in V_\Omega$$

$$S : V_{\mathcal{B}} \to V_\Omega \text{ is given by } \alpha(gS) = \sum_{\substack{B \in \mathcal{B} \\ (\alpha, B) \in I}} Bg \qquad \text{where } \alpha \in \Omega,\ g \in V_{\mathcal{B}}$$

It is clear that V_Ω, $V_{\mathcal{B}}$, S and T completely determine the structure. It is also clear that $\dim V_\Omega = v$ and $\dim V_{\mathcal{B}} = b$. Since T and S are both linear transformations we see that if T is a monomorphism then $\dim V_\Omega \leq$

dim $V_{\mathcal{B}}$ and so $v \leq b$. Since V_Ω is a space of functions there exists a linear transformation $J : V_\Omega \to V_\Omega$ such that $(\alpha) fJ = \sum_{\beta \in \Omega} \beta f$. We could define a similar map for $V_{\mathcal{B}}$ but there seems in our applications to be no need of it.

<u>5.10 Proposition.</u> Let $(\Omega, \mathcal{B}, I)$ be a $t-(v, k, \lambda)$ structure. Then $TS = (r - \lambda_2) I + \lambda_2 J$. Thus T is a monomorphism and $v \leq b$ (when K has characteristic 0).

<u>Proof.</u> Let $f \in V_\Omega$ and consider $\alpha(fTS)$.

$$(\alpha f T) S = \sum_{\substack{B \in \mathcal{B} \\ (\alpha, B) \in I}} (B)(fT)$$

$$= \sum_{\substack{B \in \mathcal{B} \\ (\alpha, B) \in I}} \sum_{\substack{\gamma \in \Omega \\ (\gamma, B) \in I}} \gamma f .$$

If $\gamma \neq \alpha$ there are λ_2 blocks B such that $\{\alpha, \gamma\} IB$. If $\gamma = \alpha$ there are r blocks B such that $\{\alpha\} IB$, because $(\Omega, \mathcal{B}, I)$ is a $t-(v, k, \lambda)$ structure. Thus

$$(\alpha f) TS = r \, . \, \alpha f + \sum_{\gamma \neq \alpha} \lambda_2 \gamma f .$$

So

$$TS = (r - \lambda_2) I + \lambda_2 J .$$

Let $f \in \ker T$. Then $fTS = 0$ and so $(\alpha)\{(r - \lambda_2 f\} = -\lambda_2 c$ where $c = \sum_\gamma \gamma f$. Thus f is constant and it is straightforward to check that f is zero.

So $\ker T = 0$ and the result now follows. //

<u>5.11 Corollary.</u> If $t \geq 1$ then $r \geq k$.

<u>Proof.</u> This follows at once by (5.9(b)). //

We will now give some examples of $2-(v, k, \lambda)$ designs constructed from vector spaces over a finite field. Let F be a finite field of order q and let V be a vector space over F of dimension n.

Choose Ω to be the set of 1-dimensional subspaces of V and choose for $\mathcal{B}$ the set of d-dimensional subspaces of V. If incidence is defined as inclusion we obtain a $2-(v, k, \lambda)$ design where

$$v = (q^n - 1)/(q - 1),$$
$$k = (q^d - 1)/(q - 1) \quad \text{and}$$
$$\lambda = \frac{(q^{n-2} - 1)(q^{n-3} - 1) \ldots (q^{n-d+1} - 1)}{(q^{d-2} - 1)(q^{d-3} - 1) \ldots (q - 1)} .$$

The proof of this is an exercise in counting numbers of subspaces and is left to the reader. If $n = 3$ and $d = 2$ then the design has parameters $2-(q^2 + q + 1, q + 1, 1)$ and $v = b$ and $r = k$. Further any two blocks have exactly one point in common. A $2-(v, k, 1)$ design with $k \geq 3$ satisfying the condition that any two blocks have precisely one point in common is called a <u>projective plane</u>.

However our main interest will be in designs constructed in an alternative way. This time we consider families of d-dimension subspaces $\{W_i \mid 1 \leq i \leq e\}$ such that $W_i \cap W_j = 0$ if $i \neq j$ and $\cup W_i = V$. Given such a family we define Ω to be the elements of V and $\mathcal{B}$ to be the set of cosets of the W_i. Incidence is again defined by inclusion. In this way we have constructed a $2-(q^n, q^d, 1)$ design.

It is now necessary to consider the existence of such families. Since $|W_i| = q^d$, $1 \leq i \leq e$ and $W_i \cap W_j = 0$ it follows that $(q^n - 1) = e(q^d - 1)$. Thus d divides n is a necessary condition. There are a number of ways to show that this condition is sufficient for the existence of families of subspaces with the desired properties. We will sketch one of them. Let d divide n and let E be an extension of F of degree d, such an extension exists from the results in Chapter 3. Let U be a vector space of dimension $e = n/d$ over E. Let $U_1, \ldots, U_e$ be the family of one-dimensional E-subspaces of U. If we now restrict the field of scalars to be F we see that U is an n-dimensional vector space over F and that $\{u_1, \ldots, u_e\}$ is an appropriate family of d dimensional subspaces.

Thus we have the existence of a $2-(q^n, q^d, 1)$ design for all prime powers q and integers $0 < d < n$, $d|n$. The special case when n is even and $d = n/2$ is called an <u>affine plane.</u> More generally any $2-(k^2, k, 1)$ design is called an <u>affine plane.</u>

Let $\mathcal{S} = (\Omega, \mathcal{B}, I)$ be a structure. Then an automorphism A is defined to be a pair of permutations (A_1, A_2) of Ω and $\mathcal{B}$ respectively such that $(\alpha A_1, BA_2) \in I$ if and only if $(\alpha, B) \in I$, for all pairs $(\alpha, B) \in \Omega \times \mathcal{B}$. We can consider an automorphism just as a permutation $\bar{A}$ on $\Omega \cup \mathcal{B}$ given by $x\bar{A} = xA_1$ if $x \in \Omega$ and $xA = xA_2$ if $x \in \mathcal{B}$, but there seems to be little advantage in so doing.

If we consider the structure $\mathcal{S}_1$ defined earlier then the pair $A = (A_1, A_2)$ is an automorphism of $\mathcal{S}_1$ where

$iA_1 = (i+2) \pmod 6$ and

$$(B_0)A_2 = B_8, \quad (B_1)A_2 = B_9, \quad (B_2)A_2 = B_2$$
$$(B_3)A_2 = B_5, \quad (B_4)A_2 = B_0, \quad (B_5)A_2 = B_6$$
$$(B_6)A_2 = B_3, \quad (B_7)A_2 = B_1, \quad (B_8)A_2 = B_4, \quad (B_9)A_2 = B_7 .$$

Clearly the set of all automorphisms of a structure $\mathcal{S}$ is a group and this is called the automorphism group of $\mathcal{S}$, $\mathrm{Aut}(\mathcal{S})$. Such a group has an action on both points and blocks and we investigate the relationship between them. We note that, in the example just given above, if we consider the group $G = \langle A \rangle$ then the orbits of G on points are $\{0, 2, 4\}$ and $\{1, 3, 5\}$. The orbits of G on blocks are $\{B_0, B_4, B_8\}$, $\{B_1, B_7, B_9\}$ $\{B_2\}$ and $\{B_3, B_5, B_6\}$. If the relationship between these orbits is examined we find that it satisfies the conditions of the following definition.

Let $(\Omega, \mathcal{B}, I)$ be a structure. Let $d \geq 1$ and $\Omega = \Omega_1 \cup \ldots \cup \Omega_d$ and $\mathcal{B} = \mathcal{B}_1 \cup \ldots \cup \mathcal{B}_e$ be partitions of the point set and the block set respectively. If there exist natural numbers a_{ij}, $1 \leq i \leq d$, $1 \leq j \leq e$ such that for all

$B \in \mathcal{B}_j$, $|\{\alpha \mid \alpha \in \Omega_i \text{ and } \alpha I B\}| = a_{ij}$, then the decomposition is called <u>block tactical.</u>

In the above example we chose the orbits $\Omega_1 = \{0, 2, 4\}$, $\Omega_2 = \{1, 3, 5\}$ and $\mathcal{B}_1 = \{B_0, B_4, B_8\}$, $\mathcal{B}_2 = \{B_1, B_7, B_8\}$, $\mathcal{B}_3 = \{B_2\}$ and $\mathcal{B}_4 = \{B_3, B_5, B_6\}$. We have

$$a_{11} = 2,\ a_{12} = 1,\ a_{13} = 1,\ a_{14} = 1,$$
$$a_{21} = 1,\ a_{22} = 2,\ a_{23} = 0,\ a_{24} = 2\ .$$

We now come to an important technical result sometimes called Block's Lemma. Our approach is inspired by that of Cameron and Liebler [2] and we remind the reader that we will use the notation introduced just before (5.10).

<u>5.12 Lemma.</u> Let $(\Omega, \mathcal{B}, I)$ be a structure such that T is a monomorphism. Let $\Omega = \Omega_1 \cup \ldots \cup \Omega_d$ and $\mathcal{B} = \mathcal{B}_1 \cup \ldots \cup \mathcal{B}_e$ be a block tactical decomposition of $(\Omega, \mathcal{B}, I)$. Then $d \leq e$.

<u>Proof.</u> Let V_Ω and $V_\mathcal{B}$ be defined as in the remarks before (5.10). Let V^0_Ω be the subspace of V_Ω generated by the characteristic functions of the Ω_i $(1 \leq i \leq d)$ and $V^0_\mathcal{B}$ be the subspace of $V_\mathcal{B}$ generated by the characteristic function of the $\mathcal{B}_i$ $(1 \leq i \leq e)$.

Note that $\dim V^0_\Omega = d$ and $\dim V^0_\mathcal{B} = e$. Since T is injective (5.10), if we can show that $(V^0_\Omega)T \leq V^0_\mathcal{B}$ the result follows. However this is precisely the implication of being block tactical since, if we denote characteristic function on a set U by χ_U we note $(\chi_{\Omega_j})T = \Sigma a_{ij}\chi_{\mathcal{B}_j}$. //

<u>5.13 Theorem.</u> Let G be a subgroup of the automorphism group of $t-(v, k, \lambda)$ structure. Assume that in its action on points G has d orbits and that in its action on blocks it has e orbits. Then $d \leq e$.

<u>Proof.</u> We know from (5.10) that in this situation T is a monomorphism. However when we examine the orbits as partitions it is easy to see that they form a block tactical decomposition. The result now follows from (5.12). //

5.14 Corollary. If G is a subgroup of the automorphism group of a $t-(v, k, \lambda)$ structure which is transitive in its action on blocks then it is transitive in its action on points. //

Given any structure (Ω, B, I) a flag is defined to be an incident point-block pair. It is worth pointing out that U. Ott [20] has constructed an algebra based on the greater space $V_{\mathfrak{F}}$ of mappings $\mathfrak{F} \to K$, where $\mathfrak{F}$ is the set of flags and K is some field of characteristic zero. He then defines two endomorphisms X and Y of $V_{\mathfrak{F}}$ substantially as follows: for $f \in V_{\mathfrak{F}}$ and $(\alpha, B) \in \mathfrak{F}$

$$((\alpha, B)f)X = \sum_{\substack{C \\ (\alpha,C)\in I}} (\alpha, C)f \quad \text{and}$$

$$((\alpha, B)f)Y = \sum_{\substack{\beta \\ (\beta,B)\in I}} (\beta, B)f \;.$$

The advantage of this approach is that if we consider the algebra generated by X and Y inside $\mathrm{End}_K(V_{\mathfrak{F}})$ we have an algebra which contains all the incidence information about the structure (provided it is unital).

The mappings $i_1 : V_\Omega \to V_{\mathfrak{F}}$ defined by $(\alpha, B)gi_1 = \alpha g$, $g \in V_\Omega$ and $i_2 : V_{\mathcal{B}} \to V_{\mathfrak{F}}$ defined by $(\alpha, B)gi_2 = Bg$, $g \in V_{\mathcal{B}}$ are respectively embeddings of V_Ω in $V_{\mathfrak{F}}$ and of $V_{\mathcal{B}}$ in $V_{\mathfrak{F}}$. It is straightforward to check that $Ti_2 = i_1Y$ and $Si_1 = i_2X$. It is therefore possible to embed our arguments into the algebra language of Ott. A disadvantage of this approach is that it seems to make the arguments much messier, but given the results which Ott has been able to prove using this algebra it is important to point out the connection. In fact Ott defines X and Y slightly differently: his X is our $X - \mathrm{Id}$ and his Y is our $Y - \mathrm{Id}$, but the algebra obtained is the same.

An automorphism group of a structure clearly induces a permutation group on the set of flags. It is clear that transitivity on flags implies both transitivity on blocks and on points. However in general it is a much stronger condition and it is known that in many cases transitivity on flags implies that the action is primitive on points (see Chapter 1, p. 5). This condition is important in the next section and enables us to use Hering's results to study the automorphism groups of various designs. We first need a combinatorial lemma.

5.15 Lemma. Let $\mathcal{S}_1$ be a $2-(v_1, k_1, \lambda)$ structure and $\mathcal{S}_2$ be a $2-(v_2, k_2, \lambda)$ structure such that $v_2 = ev_1$ for some integer $e > 1$. Suppose that r_i, $i = 1, 2$, is the number of blocks of $\mathcal{S}_i$ incident with a given point of $\mathcal{S}_i$, and that $r_1 = r_2 = r$. Then $(r, \lambda)((r, \lambda) - 1) \geq \lambda$.

Proof. Let $d = (r, \lambda)$. Using the equations of (5.9) for $\mathcal{S}_1$ and $\mathcal{S}_2$ respectively we obtain

$$\lambda(ev_1 - 1) = r(k_2 - 1) \qquad \text{and}$$

$$e\lambda(v_1 - 1) = er(k_1 - 1).$$

Thus

$$\lambda(e - 1) = r(k_2 - 1 - ek_1 + e).$$

So

$$\lambda(e - 1) = r(e - 1 - (ek_1 - k_2)). \qquad (*)$$

Also

$$\lambda = r\{1 - \frac{ek_1 - k_2}{e - 1}\}.$$

Since $\mathcal{S}_1$ is a $2-(v_1, k_1, \lambda)$ structure, $r > \lambda$ and hence

$$0 < 1 - \frac{ek_1 - k_2}{e - 1} < 1.$$

So

$$0 < ek_1 - k_2 < e - 1.$$

Writing $\lambda = dh_0$ where $(h_0, r) = 1$, $(*)$ gives $h_0(e - 1) = (r/d)(e-1-(ek_1 - k_2))$.

So

$$d(e - 1) = r\{\frac{e-1-(ek_1 - k_2)}{h_0}\} = rh$$

say where h is a positive integer.

Using the inequality (5.10) for $\mathcal{S}_2$, $r \geq k_2$, and so

$$d = \frac{rh}{e-1} \geq \frac{k_2 h}{e-1} = \frac{ek_1 h}{e-1} - \frac{(ek_1 - k_2)h}{e-1} .$$

But

$$-\{\frac{ek_1 - k_2}{e-1}\} > -1$$

and so

$$d > h\{\frac{ek_1}{e-1} - 1\} > k_1 - 1 .$$

Since $\lambda(v_1 - 1) = rk_1$ (5.9), (r/d) divides $(v_1 - 1)$. Thus $\frac{(v_1-1)d}{r} \geq 1$ and so $\lambda\{\frac{(v_1-1)d}{r}\} \geq \lambda$. Hence from (5.9(c))

$$(k_1 - 1)d \geq \lambda .$$

But $d > k_1 - 1$, so $d-1 \geq k_1-1$, and hence $(d-1)d \geq \lambda$ as required. //

5.16 Theorem. Let $\mathcal{S}$ be a $2-(v, k, \lambda)$ structure such that $\lambda \geq d(d-1)$ where $d = (r, \lambda)$. Let G be a subgroup of $\mathrm{Aut}(\mathcal{S})$ and assume that G acts as a transitive permutation group on the flags. Then G is primitive in its action on the points.

Proof. Consider the action of G on the points Ω. Assume that G is not primitive in this action. Let Ω_1 be a set of imprimitivity for G, i.e. $\Omega_1 \subset \Omega$, $|\Omega_1| > 1$ and if $g \in G$ $\Omega_1 g \cap \Omega_1 = \emptyset$ or Ω_1. We now define a new structure $\mathcal{S}_1$ whose point set is Ω and whose blocks are those blocks B of $\mathcal{S}$ such that there exists α in Ω_1 so that $\alpha I B$. Incidence is the restriction of the incidence of $\mathcal{S}$.

Because G is transitive on flags it follows that $\mathcal{S}_1$ is a $2-(v_1, k_1, \lambda)$ design where $v = ev_1$ and $e = |\Omega| / |\Omega_1|$. Clearly the number of blocks incident with $\alpha \in \Omega_1$ is the same in either structure so we can apply the preceding lemma. The result then follows. //

5.17 Corollary. Let $\mathcal{D}$ be a $2-(v, k, 1)$ design and let $G \leq \mathrm{Aut}(\mathcal{D})$ so that G is transitive on flags. Then G is primitive in its action on points. //

Note that a $2-(v, k, 1)$ structure is automatically a design. We end the section by giving some examples of $2-(v, k, 1)$ designs which have flag-transitive automorphism groups. Let E be a field of order q^n, where q is a prime power.

Given any $a \in E$, $x \in U(E)$ and r in the Galois group H of E over its prime field, we define a mapping $T(a, x, \sigma)$ $E \to E$ by $uT(a, x, \sigma) = xu^{\sigma} + a$ $\forall u \in E$. The set of all such mapping is called the affine group of semilinear mappings $A\Gamma L(1, q^n)$. The set $\{T(a, 1, 1) : a \in E\}$ is a normal subgroup of $A\Gamma L$ called the translation group and the set $\{T(0, x, \sigma) : x \in U(E), \sigma \in H\}$ is called the group of semilinear mappings of E and denoted by $\Gamma L(1, q^n)$. This group is a soluble group of order $q^n(q^n - 1)n$.

Choose a subfield F of E of order q^d, $d|n$. If we look at the orbit of F under the action of $T(a, x, 1)$ it can be seen that this set of subsets of E gives rise to a $2-(q^n, q^d, 1)$ design whose points are E and whose blocks are the images of F under the group. This is a different representation of the $2-(q^n, q^d, 1)$ designs which were constructed earlier. It is, however, easier to see that these designs have flag-transitive automorphism groups.

There are a number of ways of showing this, and we will sketch one method. We first observe that when G is a flag-transitive automorphism group of a $2-(v, k, \lambda)$ design and $\alpha \in \Omega$ then G_{α} is transitive on the set of blocks incident with α.

5.18 Lemma. Let $\mathcal{S} = (\Omega, \mathcal{B}, I)$ be a $2-(v, k, \lambda)$ structure and let $G \leq \mathrm{Aut}(\mathcal{S})$. Assume that for each $\alpha \in \Omega$ G_{α} acts transitively on the blocks incident with α. Then G is flag-transitive.

Proof. We first show that G is transitive on blocks. Let $B_1, B_2 \in \mathcal{B}$. If there exists $\alpha I B_1$ and $\alpha I B_2$ then $\exists g \in G$ such that $B_1 g = B_2$ by hypothesis. Otherwise choose γ such that $\gamma I B_1$ and γ is not incident with B_2. Choose $\delta I B_2$, and B_3 a block such that $\gamma I B_3$ and $\delta I B_3$. (This is possible since $t = 2$.) Now $\exists g_1$ such that $B_1 g_1 = B_3$ (using G_γ) and $\exists g_2$ such that $B_3 g_3 = B_2$ (using G_δ). Thus G is transitive on blocks. Thus G is transitive on points (5.14). Now let (α, B) (γ, C) be any pair of flags. There exists $g \in G$ such that $\alpha g = \gamma$ and $h \in G_\gamma$ such that $Bgh = C$ as required. //

Using this lemma it is a routine calculation to check that the automorphism group $T(\alpha, x, 1)$ of E induces a flag transitive automorphism group of the $2 - (q^n, q^d, 1)$-design.

APPLICATIONS OF HERING'S THEOREM

The main applications of the theorem to be discussed concern primitive permutation groups. Let G be a finite primitive permutation group on a set Ω. Let G have a normal abelian subgroup. Then by (1.9) A is an elementary Abelian normal subgroup of G and we can identify A with Ω. If $\alpha \in \Omega$ $G = A \cdot G_\alpha$ and G_α is a subgroup of $\mathrm{Aut}(A)$. Since A is an elementary Abelian p-group, A is a vector space over $GF(p)$ of dimension say n, where $|A| = p^n$. Thus $G_\alpha \leq GL(n, p)$. As a consequence the structure of G is now a problem in the theory of linear groups.

The first specific example we look at is when G is a doubly transitive group. Recall from Chapter 1 that a permutation group is said to be transitive if G has just one orbit on Ω.

We now define n-transitivity recursively by defining 1-transitivity to be transitivity and by saying that G is n-transitive if G_α is $n - 1$ transitive on $\Omega - \{\alpha\}$ for each α in Ω. This is equivalent to saying that G is transitive on the set of ordered subsets $\{\alpha_1, \ldots, \alpha_n\}$ (where $\alpha_i \in \Omega$ and $\alpha_i \neq \alpha_j$ if $i \neq j$) where $g \in G$ acts on $\{\alpha_1, \ldots, \alpha_n\}$ by mapping $\{\alpha_1, \ldots, \alpha_n\} \to \{\alpha_1 g, \ldots, \alpha_n g\}$. Clearly n-transitivity implies k-transitivity for $n > k$. Assume that G is 2-transitive on Ω and that $\Delta \subset \Omega$ be a set of imprimitivity. Let $\alpha, \beta \in \Delta$ and $\gamma \notin \Delta$. Then there is g in G such that $(\alpha, \beta)g = (\alpha, \gamma)$. Then $g \in G_\alpha$ and $\Delta g \neq \Delta$. This cannot happen, so if G is 2-transitive then G is primitive. Also since G_α is transitive on $\Omega - \{\alpha\}$ we see that $n - 1$ divides $|G_\alpha|$.

Let V be $GF(p^a)$ and recall that $A\Gamma L(1, p^a)$ is the set of mappings $\{x \to ax^\sigma + b \mid a \in V^\#, b \in V \text{ and } \sigma \in Gal(V/GF(p))\}$. Now $A\Gamma L(1, p^a)$ contains the normal subgroup $\{x \to x + b \mid b \in V\}$ which is isomorphic to V as an additive group, and is known as the translation group. We will write the group also as V. Now $A\Gamma L(1, p^a)/V$ is isomorphic to $\Gamma L(1, p^a)$.

However $A\Gamma L(1, p^a)$ is always doubly transitive since V is clearly transitive on V (by addition) and $\Gamma L(1, p^a)$ is transitive on $V - \{0\}$ by multiplication. Note that as a permutation group on V, $\Gamma L(1, p^a)$ is the stabiliser of 0 in $A\Gamma L(1, p^a)$, which we have seen is soluble of order $p^a . (p^a - 1) . a$. The next theorem (due originally to B. Huppert, see [16]) says that apart from a finite number of known exceptions subgroups of $A\Gamma L(1, p^a)$ exhaust the list of soluble 2-transitive groups.

5.19 Theorem. Let G be a soluble 2-transitive group acting on a set Ω where $|\Omega| = n$. Then $n = p^a$ for some prime and $G \subseteq A\Gamma L(1, p^a)$ unless $p^a = 3^2, 5^2, 11^2, 23^2$, and 3^4.

Proof. Since G is primitive and soluble we know that G contains an Abelian normal subgroup, and that if $\alpha \in \Omega$, $G_\alpha \subseteq GL(a, p)$ for some prime p and integer $a \geq 1$. Since G is doubly transitive, G_α is transitive on $\Omega - \{\alpha\}$ so that $(p^a - 1) \,\big|\, |G_\alpha|$ and $\phi_a^*(p)$ divides $|G_\alpha|$. Hence by Theorem 5.7 the result almost follows. The cases when $n = 2^6$ and when $n = q^2$, $q = 2^3 + 1$ for some j are both of either small order or of small degree and we omit further details, which can be found in [16]. //

Clearly if G is not soluble but contains a normal Abelian subgroup the results of Theorem 5.6 hold. Thus we may assume that G contains only one non-Abelian chief factor and this factor is simple and the remaining structure is given in a particularly nice way. Given the classification of finite simple groups it is then possible to classify 2-transitive groups which contain a regular normal subgroup. This task has been completed by C. Hering [13]. It is beyond the scope of these notes since it involves a considerable knowledge of the structure of finite simple groups. In particular we need to have information about the representations of such groups over finite fields.

The second application is to the situation where G is the automorphism group of a block design.

5.20 Theorem. Let $\mathfrak{D}$ be a 2 - (v, k, 1) block design where $k | v$. Let $G \leq \mathrm{Aut}(\mathfrak{D})$ such that G is flag-transitive and soluble. Then $v = p^a$ for some prime p and $a > 0$, and $G \leq A\Gamma L(1, p^a)$ unless $p^a = 3^2, 5^2, 7^2, 11^2, 23^2$ or 3^4.

Proof. By the earlier results (5.16) we know that G is primitive on the points of $\mathfrak{D}$. As G is soluble G has an Abelian regular normal subgroup and is $v = p^a$, and if α is a point $G_\alpha \leq GL(a, p)$. Since G is flag-transitive G_α is transitive on the set of blocks incident with α. As there are $(v - 1)/(k - 1)$ such blocks, $(v - 1)/(k - 1)$ divides the order of G_α. Because $k | v$, $k = p^e$ for some e and because $p^e - 1$ divides $p^a = 1$, e divides a. Then $(p^a - 1)/(p^e - 1)$ divides the order of G_α and so $\phi^*_a(p)$ divides $|G_\alpha|$. Now the results of (5.8) hold and the theorem then follows (once again after some special arguments for low dimensional cases). //

This result was first proved in the special case that $\mathfrak{D}$ is an affine plane (that is $v = k^2$) by D. A. Foulser [8]. It was realised by A. R. Camina and T. M. Gagen [3] that the proof held in much greater generality.

As was noted after (5.18) these subgroups do in fact act flag transitively on appropriate designs. However a complete classification of the subgroup which acts in this way has not yet been completed. We should add that there is rather more similarity between (5.19) and (5.20) than may appear at first sight. This is because group G is a flag-transitive automorphism group of a 2 - (v, 2, 1) design if and only if it acts 2-transitively on the set of points.

As in the 2-transitive case using Theorem 5.6 it should be possible to classify all flag transitive automorphism groups with regular normal Abelian subgroups using the classification of finite simple groups. This task needs to be carried out but has not yet been done. Further A. R. Camina and T. M. Gagen

[4] have shown that if G is a flag transitive automorphism group of a 2 - (v, k, 1) design with $k|v$ and if G has a non-Abelian minimal normal subgroup then G has a simple minimal normal subgroup. Thus it does seem that a complete classification of such automorphism groups should now be possible.

References

1. G. D. Birkhoff and H. S. Vandiver, "On the integral divisors of $a^n - b^n$", Ann. of Math., 5 (1904), 173-180.
2. P. J. Cameron and R. A. Liebler, "Tactical decompositions and orbits of projective groups", Linear Algebra and its Applications, 46 (1982), 91-102.
3. A. R. Camina and T. M. Gagen, "Finite groups with maximal subgroups of odd order", Archiv der Math., 28 (1977), 357-368.
4. A. R. Camina and T. M. Gagen, "Block transitive automorphism groups of designs", J. of Alg., 86 (1984), 549-554.
5. A. H. Clifford, "Representations induced in an invariant subgroup", Ann. of Math., 38 (1937), 533-550.
6. N. J. Divinsky, "Rings and radicals", George Allen and Unwin, London, 1965.
7. D. A. Foulser, "The flag transitive collineation groups of the finite Desarguesian affine planes", Canadian J. Math., 16 (1964), 443-472.
8. D. A. Foulser, "Solvable flag transitive affine groups", Math. Zeit., 86 (1964), 191-204.
9. A. W. Goldie, "The structure of Noetherian rings", in "Lectures on Rings and Modules", Lecture Notes in Mathematics 246, Springer Verlag, Berlin, 1972.
10. P. R. Halmos, "Finite-dimensional vector spaces" (second edition), D. Van Nostrand, New York, 1958.
11. C. Hering, "Zweifach transitive Permutations gruppen, in denen 2 die maximale Anzahl von Fixpunkten von involutionen ist", Math. Zeit., 104 (1968), 90-102.
12. C. Hering, "Transitive linear groups which contain irreducible subgroups which contain subgroups of prime order", Geometrica Dedicata, 2 (1979), 425-460.

13. C. Hering, "Transitive linear groups which contain irreducible subgroups of prime order II", to appear.

14. I. N. Herstein, "Noncommutative rings", Carus Mathematical Monographs 15, J. Wiley and Sons for the Mathematical Association of America, 1968.

15. D. R. Hughes and F. Piper, to appear, Cambridge U. P., London.

16. B. Huppert, "Zweifach transitive, auflösbare gruppen", Math. Zeit., 68 (1957), 126-150.

17. B. Huppert, "Endliche Gruppen I", Springer Verlag, Berlin, 1967.

18. B. Huppert, "Lineare auflösbare Gruppen", Math. Zeit., 67 (1957), 479-518.

19. H. Lüneburg, "Ein einfacher Beweis für den Satz von Zsigmondy über primitive Primteiler von $A^N - 1$", in "Geometries and Groups", ed. M. Aigner and D. Jungnickel, Springer Verlag, Berlin, 1981.

20. U. Ott, "Some remarks on representation theory in finite geometry", in "Geometries and Groups", ed. M. Aigner and D. Jungnickel, Springer Verlag, Berlin, 1981.

21. U. Ott, "Bericht über Hecke Algebren und Coxeter Algebren endlicher Geometrien", in "Finite Geometries and Designs", ed. P. J. Cameron, J. W. P. Hirschfeld and D. R. Hughes, Cambridge U. P., London, 1981.

22. K. Zsigmondy, "Zur Theorie der Potenzreste", Monatshefte für Math. und Phys., 3 (1892), 265-284.

The following books are not cited in the text, but are good references for the background material we have assumed. The list is not intended to be exhaustive.

A. A. W. Chatters and C. R. Hajarnavis, "Rings with chain conditions", Pitman, London, 1980.

B. P. M. Cohn, "Algebra", Volume 1 (second edition), J. Wiley and Sons, Chichester, 1982.

C. B. Hartley and T. O. Hawkes, "Rings, modules and linear algebra", Chapman and Hall, London, 1970.

D. I. N. Herstein, "Topics in algebra", Xerox College Publishing, Lexington, Mass., 1964.

E. J. S. Rose, "A course on group theory", Cambridge U. P., London, 1978.

Index